SpringerBriefs in Computer Science

T0215346

For further volumes:
http://www.springer.com/series/10028

Piyushimita (Vonu) Thakuriah
D. Glenn Geers

Transportation and Information

Trends in Technology and Policy

 Springer

Piyushimita (Vonu) Thakuriah
University of Glasgow
Glasgow
UK

D. Glenn Geers
National ICT Australia
Kensington, NSW
Australia

ISSN 2191-5768   ISSN 2191-5776   (electronic)
ISBN 978-1-4614-7128-8   ISBN 978-1-4614-7129-5   (eBook)
DOI 10.1007/978-1-4614-7129-5
Springer New York Heidelberg Dordrecht London

Library of Congress Control Number: 2013935480

Printed on acid-free paper

Springer is part of Springer Science+Business Media (www.springer.com)

To the memory of our fathers
Bhabesh Chandra Thakuriah
Donald Leslie Geers

Preface

In this book, we review recent developments at the intersection (no pun intended) of Information and Communications Technology (ICT) and surface transportation, and the technical, social and institutional challenges stimulated by these trends. Developments in pervasive sensing and widespread proliferation of vast numbers of mobile and static sensors promise to bring a sea-change in the way transportation information can be designed and used, particularly with Machine-to-Machine communications and by information generated by people-centric sensors. Methods to manage and analyze such data have led to novel mobility services which may have the potential to lead to sustainable and socially interesting travel. The use of such information stimulates numerous social, institutional, ethical, and legal challenges, some of which we have attempted to bring together in this book.

The book is aimed at researchers, graduate students, industry professionals, and decision-makers considering problems in surface transportation and approaches and limitations of ICT in understanding and addressing these problems. The use of the word transportation throughout the book refers to surface transportation, unless explicitly noted otherwise. Contributions from a number of academic disciplines have made these myriad developments possible. We take a broad-based view of policy and the underlying organizing theme is one of economic, environmental, and social sustainability. It is our hope that we have been able to bring together this broad spectrum of knowledge in this brief volume, albeit in a limited way. Our eventual goal is awareness-building about a wide range of problems in ICT and transportation, thereby stimulating research approaches that address multiple concerns and perspectives. The book is broad and non-technical in nature, with an emphasis on being a survey as opposed to an exhaustive treatment of a small set of topics. It may be read as part of an introduction to a graduate course on transportation and technology offered in transportation planning, transportation engineering, computer science, geography, or public administration.

The book begins, in Chap. 1, with an overview of the many facets of ICT in transportation, including Intelligent Transportation Systems (ITS), Location-Based Services (LBS), relevant aspects of smart and connected cities, dynamic resource management, mobile health, and assistive technologies. We also discuss environmental, economic, and social sustainability outcomes which an information-centered mobility environment can potentially address. In Chap. 2 we present an overview of the major existing and emerging sensor and communications technologies and describe the types of information they generate. Chapter 3 follows with a range of systems and services that utilize these sources of information. Chapter 4 addresses institutional, legal, and coordination issues as well as issues of behavioral effects and societal preparedness to handle the information-centered mobility environment. Conclusions and possible future directions are given in Chap. 5.

Whereas ITS and LBS have been very active research areas in transportation, the contributions of ICT have been greater than solutions and services developed under such banner. Examples include strategies for mobility-on-demand, mobility assistance for persons with disabilities, smart cities and ubiquitous information environment, community and urban informatics, resource management and asset condition monitoring. Although we try to devote space to many different types of ICT examples in transportation, we had to be selective, thereby making greater discussions of certain concepts than others and the book is far from an exhaustive survey of all that has been done on this vast topic. By a survey, we also mean that we do not go into detailed discussion of any one topic and attempt to merely provide an overview of what has been done in an area. Moreover, the emphasis is on the transportation system and service aspects and not on the details of the technology and methodological aspects.

The book was stimulated by our involvement in many research projects that are too numerous to list. However, virtually each of these projects gave us the ability to explore and appreciate the technical, social, and management challenges associated with the emerging information-centered mobility environment.

The book would not have been possible without the strong support of our spouses, Claude Hanhart and Nicole Geers, and Glenn's daughter, Céline. We would like to acknowledge the contribution of Dr. Caitlin Cottrill, University of Aberdeen, UK and John Laird, NICTA and University of New South Wales, who helped with reading through and editing the book. The book was written while the first author worked in Chicago and we would also like to thank ICT for making the 9,300-mile collaboration possible.

Glasgow and Sydney   Piyushimita (Vonu) Thakuriah
January 2013   D. Glenn Geers

Contents

Chapter 1
Introduction

1.1 Trends in ICT-Based Surface Transportation

This book is concerned with the use of Information and Communications Technology (ICT) in the field of surface transportation. ICT is a major driver of both economic growth and improved quality of life in the new global economy. ICT has evolved over centuries and innovations in ICT have occurred throughout history. This book is concerned with digital ICT, which has its basis in computer software, hardware and communications systems, and the explosive development of which spans the last fifty or so years of human history.

Surface transportation has historically been a fundamental backbone of economies and societies, by contributing significantly to the Gross Domestic Product (GDP), employment and overall support of trade and commerce. Surface transportation is also a critical ingredient in the quality of life by enabling travel to jobs, educational, social and recreational activities. Surface transportation systems can also lead to poor economic and social outcomes because of traffic congestion, road fatalities, air pollution, Green House Gas (GHG) emissions and continued dependence on fossil fuels. Transportation technology, like ICT, has greatly evolved over time from primarily muscle (human or animal) powered systems to current-day motorized mobility for passengers and freight.

ICT has the potential to contribute to environmentally sustainable and safe travel and mobility management. Some examples of how such outcomes can result include:
Environmental Sustainability: On-board diagnostics of environmental pollution in cars, road traffic signals and motorway ramp meters which lead to shorter vehicle idling time, vehicle engines that turn off automatically when the vehicle is stationary and automatic eco-feedback to inform drivers about the environmental impacts of their driving behavior may enable eco-friendly travel;
Safety: Sensors which remotely telemonitor the health of a driver, anticipatory weather information to help drivers avoid hazardous driving conditions, cars that brake without driver intervention when sensing an obstruction ahead or warn drivers when they are overly fatigued or distracted may enable safer travel;

P. (Vonu) Thakuriah and D. G. Geers, *Transportation and Information*, SpringerBriefs in Computer Science, DOI: 10.1007/978-1-4614-7129-5_1, © The Author(s) 2013

Shared Mobility and Social Transportation Systems: Social media that allow people to find a real-time walking buddy from an unsafe train station in real-time, volunteer to be a driver for seniors in the neighborhood, or to share interesting location-based information to others nearby may lead to new models of co-production of shared transportation and mobility services;

Assistive Travel: Robotic assistive technologies for persons with disabilities, scooters that follow a senior person back home with the shopping load, and augmented real-world environments in which wayfinding and navigation are made easier may assist the mobility of those with special mobility needs;

Asset and Resource Management: Bridges that autonomously report their own structural health conditions to engineers and systems that monitor and adjust the charging load of electric vehicles in accord with location-specific load on the electricity grid may enable efficient ways of managing resources.

The above examples demonstrate that ICT may be used for transportation and mobility services in diverse ways and in different application areas. In the following chapters the focus is on growing ICT-based mobility strategies and a discussion of the technical, social, policy and user-related questions in the emerging information-centered mobility environment.

As in many other ICT-rich sectors such as health, energy, defense and finance, the use of ICT in transport enables knowledge discovery and service development that would otherwise not be possible. In each of the above examples, there are questions of coupling together sensors and communications systems; methods to extract, analyze and distribute information; and issues relating to user acceptance, legal implications and management. Yet, research, development and practice at the different stages of the technology lifecycle traditionally lie with professionals from different disciplines, and feedback and learning from experiences at each stage is not automatic (for example, professionals who are involved with the evaluation of the impact of such technologies on the behavior of travelers are often not involved at the design stage, or, technology designers may not be interested in the broader societal or economic impacts of a particular sensor, communication system or mobile application). In fact, it may be quite difficult to estimate what the broader impacts would be. By focusing on the "technology and policy" aspects of transportation and ICT together, we discuss the trends relating to these complex questions and directions towards which they are headed.

1.2 Overview of the State of Information-Based Mobility Environment

The convergence of several heterogeneous technologies has made the current state of information-centered mobility a possibility. The most relevant developments are in the fields of: (1) sensors; (2) location and positioning systems; (3) information extraction technologies; (4) sensor fusion technologies; (5) communication methods; (6) information and data management systems; (7) methods for information analysis; and (8) methods to understand user dynamics and impacts associated with the use of mobility information and services.

Sensor systems collect operational details on transportation conditions and provide real-time data on current conditions for immediate service delivery and informed decision-making. The transportation sector has a vast range of specialized infrastructure-based sensors for the detection and surveillance of mobility patterns and infrastructure conditions. In-vehicle sensors in the powertrain, chassis and the body of vehicles allow myriad automated tasks ranging from monitoring energy use to vehicle handling and safety as well as situational awareness regarding hazardous conditions on the road around the driver. People-centric sensors such as the microblogs, question-and-answer databases and mobile connected devices such as cell phones with location-aware technologies promise to allow large-scale, pervasive and distributed sensing system. Wearable biometric and other person-based sensors may have applications in mobile health and wellness informatics. In virtually all of the technologies discussed in this book the ability to more-or-less precisely locate a user or an asset, is a central ingredient. Satellite-based positioning systems such as the Global Positioning System (GPS) have been transformational in strengthening the role that mobile sensors can play.

Developments in spatial data management, geographic information retrieval, information extraction to retrieve intelligence from raw data (for example, data from digital images, text, audio data) and traffic and transportation engineering and planning have increased the overall utility of raw, sensor-based information. Mobility analytics generate knowledge for functions as diverse as automatic incident detection on roads (due to vehicle collisions or lane closures due to a hazardous material spill), prediction of bus arrival times at specific locations, future traffic conditions, location-based information-sharing among members of a social network, real-time management of vehicle fleets, or information on when people should travel. Such intelligence is necessary in order to support broader economic, environmental and societal outcomes as well as for high quality personal travel and freight movements.

Currently, different entities collect data on different aspects of the transportation system in a region. Mobility analytics based on such heterogeneous data sources will be critical components of intelligent cities of the future. We see future mobility intelligence giving rise to a *Digital Mobility Information Infrastructure* (DMII), which will comprise three tiers of information sources.

Primary Tier: The primary tier will consist of a myriad of infrastructure-based, vehicle-based, mobile, portable and wearable sensors, communications and information processing elements that are briefly discussed above, and which are the main subjects of this book. Some of these sensors may not be traditionally used in the transportation sector, e.g., weather sensors, but the information they offer may be critical for quality mobility information. This tier defines a continuum of information on transportation.

Secondary Tier: The second tier of information is comprised of information that is not directly generated by the primary tier but which may support its use in various ways. Examples include maps and Points of Interest (POI) databases which are critical for advanced and flexible mobility services, together with extant data sources pertaining to the overall state of the transportation network and the mobility environment. The latter sources include synthetic, model-derived information from regional transportation planning and travel forecasting models, which have historically served as the basis for making informed decisions about transportation infrastructure investments (for example, building highways or metro rail systems). Household travel survey programs and administrative data on transportation operations and performance management also support mobility intelligence and are rightfully part of this Tier.

Tertiary Tier: The third tier is comprised of "background data" from censuses, socioeconomic and demographic data programs, along with information sources on crime, health, safety, weather, emergencies, special events and related activities, to the extent that they can assist in mobility analytics.

Data from the three tiers of the DMII will support novel applications, services and planning activities that will be enabled through a range of technologies including, but not limited to, web services, standalone applications and peer-to-peer distributed applications, depending on end-user need. Although any specific end use may involve only a small part, perhaps even a tiny part, of the overall information infrastructure, the tiers of the DMII, taken together, can potentially stimulate discoveries about mobility intelligence to support various outcomes. Outcomes include safety, environmental and economic sustainability, social justice and cost-effectiveness. Additional moblity outcomes include personal satisfaction, interestingness and social-relevancy. These are outcomes which mobility intelligence can support, if designed adequately and inserted appropriately into society.

Several factors will jointly determine the extent to which these outcomes will result: the quality of the data and the communications aspects of the information infrastructure, the accuracy of mobility analytics which use the data to support mobility intelligence, and institutional and social factors in using that information. Data quality and accuracy and the way the data are used and connected to needs is critical; more data does not always mean better information. Mobility analytics consists of data management tools, systems and simulators that are able to extract mobility-specific intelligence from the large volumes of data in order to expose system behaviour and support services. Mobility analytics also consists of operations, management and planning tools to forecast demand, choices and preferences, manage resources, and to make impact and risk assessments. The reliability and accuracy of these tools is important to unleash the full power of such an information infrastructure and its eventual connection to outcomes of interest.

For the adoption and continued participation by users in sensing technology over time, many other factors are equally important to consider at the design stage. These include: (1) economic factors such as incentives and user benefits derived (for example, travel time saved, reductions in out-of-pocket travel expenses); (2) human-computer interaction factors such as user-centered and context-aware designs that are sensitive to ease of use and to attitudes and norms of users and expectations regarding privacy and information security; and (3) social factors such as value systems satisfied, social networking opportunities enabled and the psychological benefits derived from the interestingness of resulting mobility services.

The increasingly pervasive nature of location-aware sensing environments requires that various approaches should be in place to expedite the level of societal preparedness in terms of the institutional (governance, business management) and legal infrastructure. One important aspect is the extent of digital citizenship and citizen preparedness in terms of the level of awareness, and rights and responsibilities regarding the digital mobility environment. The development of an adequate evaluation framework and operational metrics for performance measurement of services and programs derived from the DMII would be important in developing technically robust and socially acceptable systems.

1.3 Key Trends Motivating ICT Use in Transportation

ICT use in transportation today is the result of both demand (market) pull and technology push factors and the co-evolution of technology and society. Market pull factors include the overall growth in population, urbanization, motorization and consequently, congestion; eco-considerations motivated by energy use, greenhouse gas emissions, and climate change; infrastructure needs relative to funding availability, capacity and age; safety, particularly crash avoidance and after-effect management; and changing demographics and special needs emerging from aging populations.

Technology push factors include developments in the areas described in Sect. 1.2: positioning, sensing, computing, communications, wireless networking technology, and methods relating to information extraction, analysis and simulation. The evolving nature of technical work in the transportation industry including the automotive industry, Original Equipment Manufacturers (OEM), the fleet management and logistics sector, and transportation asset management and traffic management sectors have also pushed innovations along certain pathways.

The major pull factors motivating the use of ICT in transportation is a complex combination of the need to reduce congestion levels and air pollution, concerns about energy use, and overall lack of accessibility and deteriorating levels of travel quality. In 2007, the total number of motor vehicles in use in the US (excluding motor-cycles) was 247.3 million, with a motorization rate of 820 per 1000 population [1]. During that same year, the Texas Transportation Institute [2] estimated delays in travel in the US to be 5.2 billion hours, with an estimated congestion cost of $126 billion (in 2009 dollars). While it is not possible to make rigorous international comparisons on these numbers, congestion is a world-wide phenomenon that is estimated to cost over $100 billion or about 0.69 % of the GDP in the US, based on estimates in [2] and about 1 % in the EU [3]. However, these costs can be much higher in certain economies, for example, in Asian regions, where congestion costs reached an estimated 4.4 % of GDP in South Korea and 6 % in the city of Bangkok [4].

According to data from the US Energy Information Administration, motor vehicle rates per 1000 people which are already very high in developed countries are expected to remain relatively steady during 2010–2020. However, the largest increases in motorization rates are expected to occur in less-developed countries. Motorization rates are widely different among less-developed countries, at an estimated 32 motor vehicles per 1,000 persons in 2007 in China, as compared with an estimated 338 vehicles per 1,000 persons in South Korea [5]. Increasing congestion and the concomitant time and fuel wasted in travel delays, especially in large urban areas, has led governments around the world to consider innovative solutions to managing traffic and congestion. Concurrent concerns regarding energy use and cost have only added to this motivation.

Reducing road fatalities has been a major incentive for seeking technological solutions. Although data from 31 countries showed that the average annual reduction in the number of deaths between 2000 and 2009 was higher than in the three preceding decades [6], the number of deaths remain high; examples are 33,808 in the US (or 11.1 per 100,000 population), 5,772 in Japan (4.5 per 100,000 population), 7,364 in Argentina (18.4 per 100,000 population) and 6,745 in Malaysia (23.8 per 100,000 population).

Different regions have very different needs regarding infrastructure management, including the need to build new transportation facilities to address demand or to maintain existing ones due to considerations of infrastructure aging and deterioration. The result has been to either seek information-based solutions when building new facilities, e.g., by instrumenting new highways with sensors or with electronic toll collection, or in the maintenance and repair stages, with workzone-based hazard warning and information systems for motorists.

The need to prepare for natural and man-made disasters and manage emergencies has revealed the need for enhanced emergency call services that upgrade current telephone-based capabilities to text, data, images and video messages. Changing demographics have been drivers in seeking technology-based mobility strategies, for example, in nations with high rates of aging population. Policy activities in the disability community are also leading to advancements in mobility solutions through universal design considerations.

The trends toward an information-based mobility environment have also been stimulated by technology push factors. Significant developments in sensor technology have led to smart commodities ranging from household appliances to smart buildings leading to cost-efficiencies and energy savings. It is not surprising that the automotive industry has been a strong participant in this overall trend, as have been transportation management agencies and the transportation industry overall.

The increasing proliferation of personal computers, the Internet, and wireless mobile devices have opened up possibilities for information management and sharing to a previously unknown degree. Strategies by which individuals learn about, receive and use such information have changed greatly over time, reflecting the ways in which the overall digital environment has evolved as broadband speeds have increased, with 133 million high-speed lines with more than 200 kB per second in at least one direction by the end of 2009 in the US, up from 2.8 million high-speed lines in December 1999 [7], and subscriptions to mobile, wireless phone services having grown to 302.8 million, up from 207.9 million in 2005 [8].

Worldwide, in high-income countries the number of mobile phone subscribers per 100 persons is 111.07, Internet users per 100 persons is 72.20 and broadband subscribers per 100 persons is 25.76 [9]. These numbers for middle income countries are 66.63, 20.73 and 4.03, and for low-income countries, 25.07, 2.57 and 0.04, respectively. Mobile phone subscription rates are higher in each type of country compared to other modes of communication. These trends open up the opportunity for a variety of new mobility services to be developed and used on a large scale.

1.4 Mobility Technology Areas

The types of mobility needs that people and society as a whole have are diverse and the range of functions and services that industries and government agencies provide to meet these needs are enormous. The efficient movement of persons and goods from point A to point B is a complex undertaking in any society; when factors such as quality, reliability, safety, equity and sustainability are thrown into the mix, the endeavor becomes even more complex.

A diverse range of industries, small entrepreneurs, government agencies and non-profits are involved in the mobility enterprise. Similarly, several academic disciplines are involved, including geography, engineering, economics, urban planning, management sciences, computer science, electrical engineering, public health and others.

It is not surprising, given the diversity and complexity of the mobility enterprise, that the motivations for the coupling of information technology and transportation have different roots. At the time of writing this book, several thematic ways of organizing ICT related to transportation exist, ranging from broad society/system-wide concepts to specific programs that reflect these roots to varying degrees. Some are established commercial or government program areas, whereas others are newer and more experimental. These thematic ways of organizing ICT for mobility form a collection of *meta-areas*.

As the reader will see, there are no clear-cut boundaries between the meta-areas since many of the technologies used are common to different areas. The differences stem from the overall purpose and different historical starting points behind the formation of a meta-area (for example, commercial need to remotely track assets in an industry versus the need for improved traffic management in cities). Additionally, there are differences in the community to which primary stakeholders within a meta-area identify themselves as belonging (for example, traffic engineering versus location-based services versus logistics).

The meta-areas have also been affected by specific strands within the intellectual body of work, methods of practice and legacy systems from which they have evolved. While all the areas make use of sensing, positioning, communications and information processing technologies in some way, the unique research, business and management practices that have resulted from the functionalities and stakeholders associated with each meta-area have led to varying levels of advancements regarding these technologies within each area.

Location-Based Services: Location-Based Services (LBS) are information services that capitalize on the knowledge of, and are relevant to, the context of a mobile user's current or projected location. While the military and emergency management services have always used positioning information for different activities, the motivation behind the bulk of LBS in recent years has been to use advancements in positioning and communication technologies for commercial purposes. Examples of LBS include services in response to spatial queries such as "Where am I?" or "What's around me?" Other examples are directory assistance and service location (for example, find the nearest gas station with cheap gas) or Points of Interest locations (for example, find the social services building).

Intelligent Transportation Systems: Perhaps most closely associated with transportation systems and technologies are Intelligent Transportation Systems (ITS). ITS programs use a vast range of sensor, positioning, communications and related technologies that are specifically geared to enhance the performance of the transportation system. The major policy motivations for the programs are improved traffic management and congestion reduction, traveler information, and safety. ITS components are currently deployed to varying degrees in many areas around the world, for functions as diverse as traffic management and adaptive traffic signal control, electronic toll collection, and management of city bus fleets. The instrumentation of highways, bridges, transit stations and toll facilities has led to mobility services which utilize real-time data streams generated from these (primarily government-funded) sensor systems.

ITS programs are currently substantially broader than in the initial years, and include a variety of functionalities such as road weather management systems, cooperative intersection collision avoidance systems, emergency management and vehicle-based safety systems. Many of these developments have leveraged innovations in automotive electronic and sensing technology. Recent areas of activity include connected vehicles and cooperative ITS (C-ITS) programs where the rationale is that by allowing vehicles to communicate with each other and with the infrastructure around them, it is possible to deliver safety, efficiency and sustainability outcomes beyond those achieved by the same vehicles acting alone.

Smart Cities and Ubiquitous Information Societies: Currently, there are several initiatives relating to smart cities, e-cities, u-cities and digital cities, where transportation is part of a mix of overall strategies to ICT-based intelligence in cities. These concepts regarding cities are a part of a continuum that emphasize different functionalities and integration strategies. The technologies and functionalities targeted by these concepts are common but the motivations and ultimate focus may be different to a certain degree. The term "smart city" is championed by commercial entities and the expectation is that such smart city integration approaches will enable cross-agency efficiencies in a range of functionalities in multiple administrative sectors (such as traffic management, utilities and law enforcement) and ultimately to a wide spectrum of services (utilities, garbage disposal, emergency services, aged care, etc) in urban areas. The u-city concept focuses on similar functionalities but with a long-term focus on making information available anytime, anywhere and about any place [10]. These ideas stem from ubiquitous information technologies and pervasive computing which could potentially transition society as a whole to a ubiquitous information environment, where intelligence is embedded into everyday objects (termed everyware, see [11, 12]), and where information is available anytime, anywhere and about any place.

Intelligent Infrastructure Technologies for Transportation Asset Management and Condition Monitoring: Monitoring the health of highways, transit systems, tunnels and bridges is a critical function needed to prioritize maintenance and repair before damage becomes extensive, and to avoid catastrophic failures. Failures can occur due to loading, age, or environmental conditions such as earthquakes, severe weather and freeze-thaw cycles. Increased vulnerability may occur in the future due to the potential impact of climate change. There are well-developed procedures in place to regularly inspect and test asset conditions. Structural Health Monitoring (SHM) allows continuous and autonomous monitoring of the structural integrity and physical condition of a structure using embedded or attached sensors and minimum manual intervention, using non-destructive techniques. Successful implementation of SHM can replace schedule-based inspection and maintenance of structures by condition-based maintenance, thereby reducing life cycle costs significantly, and improving safety. SHM systems involve a combination of data acquisition by sensors and computational models of the structure. A vast range of specialized sensors are used for these purposes.

Informatics for Citizen Engagement and Participatory Sensing: This meta-area focuses on methods to improve citizen engagement and community involvement in urban transportation planning, as well as to leverage users in the co-production of data. Traditional models of public participation such as public hearings, open meetings with officials, processes of reviewing and commenting on posted planning documents and other formal models have been problematic [13]. ICT has opened up the possibility of bringing communities closer to the planning process on a continuing basis, in contrast to project or plan-specific purposes; yet much remains to be done in this area. There are several synergistic viewpoints that support the idea of a sustained and involved citizen participation and engagement in community and urban decisions using ICT [14–16]. Social media technologies have the potential to move public engagement to a realm beyond planning and public policy issues into the role of direct co-production of services and information that historically rested with governments and commercial enterprises. For example, Web 2.0 and social media technologies have assisted citizens in real-time location-based participatory sensing of urban spaces. Static government or commercially-produced map databases may be augmented by Volunteered Geographic Information (VGI), where geographic content may be added collaboratively. Community members may self-organize for mobility to safety in crisis situations using ICT, by seeking and sharing peer-to-peer information to supplement information from emergency management personnel. These, and many other examples have resulted in several models of User-Generated Content (UGC), which can ultimately serve to inform mobility intelligence.

Mobile Health and Technologies for Special Mobility Needs: Recent policy has focused on evidence from the public health and medical literature that being physically active is an important contributor to health and well-being. Active transportation policies target obesity as a precursor to many diseases and support bike-friendly urban design, and built environments that facilitate physical activity as a part of people's everyday lives. ICT has generally been leveraged to play a supportive role in health in several ways. Mobile health technology (mHealth) which uses mobile technologies for health research and healthcare delivery has become an established area within health informatics. One area where ICT has played a significant role is with the mobility well-being of seniors and persons with disabilities. A report by the World Bank estimates that by 2050, the number persons with disabilities will be between 10 and 12 % of the global population [17]. The increase in the over-65 population, when the risk of disability is the greatest, is expected to more than double over the next 40 years, climbing from 7 % in 2010 to 16 % or nearly 1.5 billion people globally by 2050. Assistive technologies are another area; these range from robotic walkers to scooters that autonomously follow a senior person home from shopping. Augmented real-world environments, using ubiquitous sensing and display systems which place virtual objects into the real world, are being developed to assist persons with disabilities become more mobile both within the home and outside.

1.5 Mobility Policy Areas

We take a broad-based view of policy in this book and the underlying organizing theme is one of sustainability. Our framework is best represented by that given in the report *Sustainable Transport: Priorities for Reform* produced by the World Bank [18], which defines three components of sustainable transport:

The *economic and financial component,* which includes issues of adequacy of transportation infrastructure funding, organizations and scale;

The *environmental and ecological component*, which is concerned with how transportation investments and mode options influence travel and land use patterns and how these in turn influence energy consumption, emissions, air and water quality and habitats;

The *social component*, which emphasizes adequate access to transportation by all segments of society.

The economic, environmental and social aspects of sustainable transportation systems are overall mutually reinforcing in the sense that achieving sustainability in one aspect would likely result in favorable outcomes for the other two. In fact, the "interdependent and mutually reinforcing pillars" of sustainable development are noted to be "economic development, social development, and environmental protection" [19]. How ICT can be related to these broad goals us articulated by the World Summit for Information Society's (WSIS) Declarations and Principles [20]:

> … to build a people-centered, inclusive and development-oriented Information Society, where everyone can create, access, utilize and share information and knowledge, enabling individuals, communities and peoples to achieve their full potential in promoting their sustainable development and improving their quality of life.

Sustainable Mobility Policy Area 1: Economic and Financial Aspects: Policies that address the economic and financial components of transportation systems require that resources be used efficiently and that assets are maintained properly [18] (p. 5). ICT has the potential to address the efficiency and cost-effectiveness of mobility in many ways. For example, SHM technologies have the potential to bring about efficiencies in the monitoring of bridges, tunnels, highways and transit facilities for rapid maintenance and construction.

ICT-based capacity management strategies such as traffic management systems and electronic tolling for congestion pricing or Vehicle Miles Traveled (VMT) taxes whereby drivers could potentially be taxed, not through a fuel tax, but on the basis of actual distance travelled can lead to effective use of transportation networks. Incentives could be used to encourage participants of sensing systems to report incidents (e.g., road crashes or transit facility conditions) that have the possibility of affecting operational capacity of the transportation system and travel quality. Technologies supporting safety in transportation are particularly likely to support this aspect of sustainable transportation due to the large economic and social costs associated with traffic crashes.

ICT has significant potential for dynamic resource management of vehicles (truck or bus fleets), for shared transportation (ride-sharing, car-sharing or bicycle-sharing programs) and in mobility management (coordinating rides to work or for social purposes). If these strategies are effective, the result would be more efficient management of existing transportation resources.

Finally, ICT-based mobility strategies support several industries that contribute to economic development. A far from complete list includes automobile manufacturers, Original Equipment Manufacturers, telematics companies and hardware manufacturers such as cell phone and PDA manufacturers, electronics and infotainment manufacturers, transportation and traffic sensor manufacturers, and positioning system manufacturers. ICT is integral to the business of LBS such as routing and navigation and travel and tourism information, as well as industries that support them (for example, developers of map databases, web services and communications companies). Private and non-profit providers of mobility services (car-sharing companies, paratransit, assistive mobility companies, toll road operators, companies in the health and wellness realm) also depend on ICT-based strategies for their work,

Sustainable Mobility Policy Area 2: Environmental Aspects: ICT has both a direct and an indirect role in supporting positive environmental outcomes. Examples of direct support are design and control of clean vehicle engines, development of fuel alternatives and emissions monitoring and inspection strategies. Indirect strategies include supporting transportation demand management strategies for green travel. Examples of green travel strategies are travel information and service coordination for the use of public or shared transportation instead of single occupant vehicles, congestion and pollution charging and variety of other techniques.

Virtually all of the examples of capacity management in the previous section (on economic aspects of sustainability), in fact, can also potentially give rise to environmentally-friendly outcomes leading to the notion of mutual interdependence and reinforcement among the three aspects of sustainability. Eco-driving and transportation and energy management to support electric and hybrid vehicles also have significant possibilities of reducing Green House Gases.

While these environmental outcomes will be very beneficial when realized on a large and sustained scale, it should also be kept in mind that ICT can encourage more travel by making travel easier and also that ICT has an environmental and carbon footprint. While identifying all the industries that constitute the ICT sector is itself difficult, by some estimates, this sector, consisting of ICT companies directly as well as energy consumption by ICT equipment, is responsible for a relatively small portion of global Green House Gases (GHG)—about 2–2.5 % [21]. However, identifying all industries that constitute the ICT sector is difficult and the share would be higher if indirect sources were to be accounted for. Additionally, fossil fuel use due to equipment and parts manufacturing, transport and eventual removal of waste produced by ICT materials at the end of what is typically a short lifecycle, may need to be included when accounting for the full ability of ICT to substitute for mobility. Eco-design principles will be useful in maximizing the green aspects of ICT-based mobility.

Sustainable Mobility Policy Area 3: Social Aspects: The social component of sustainable transportation systems emphasizes adequate access to transportation by all segments of society. One policy stream considers mechanisms for mobility that are equitable and socially just, and offers services to all segments of society, including those who are socially excluded and least mobile, such as low-income persons who cannot afford private cars, seniors, persons with disabilities and young children. Robust policies relating to the digital divide and open data, ICT strategies supporting shared or public transportation that are affordable and reliable, and assistive technologies for seniors and persons with disabilities are only a few examples by which the information-based mobility environment enhances the inclusiveness of all segments of society.

Transportation safety remains one of the major drivers of transportation policy as does the well-being of vulnerable populations during emergencies and disasters. Active transportation policy, which supports mobility strategies for health and well-being, is an area of increasing importance. Questions of privacy, information security and mobile digital citizenship whereby citizens are aware of their rights and responsibilities in the ubiquitous information-based mobile environment will be important policy areas.

1.6 Organization of the Book

The book reviews developments in technologies in ICT-based mobility systems and the institutional issues surrounding these technologies. Mobility decision support systems aid decision making at a variety of timescales ranging from milliseconds (for collision avoidance, for example) through minutes and hours (for journey planning for example) to decades (motorway planning, for example). In this book the scope is limited to describing existing and emerging sensors and ICT systems which aid short and medium-term mobility decision-making. The decision-maker can be a government agency, a private company, a neighborhood organization or an individual. Little emphasis is placed on the details of the methods, analysis and predictive models by which data from the sensors are used to enable transportation services. We have avoided speculating on technologies that are likely to emerge far in the future.

The rest of the book is organized in four chapters. In Chap. 2, we discuss developments in transportation sensors, communications and information management. Chapter 3 discusses several systems that use the sensors and communications systems described in Chap. 2 and collectively give rise to intelligence in transportation systems. Chapter 4 discusses institutional and sociotechnical aspects of ICT in transportation. Chapter 5 highlights major trends and summarizes the major research and implementation challenges associated with ICT-based mobility.

Chapter 2
Data Sources and Management

2.1 Introduction

For the last twenty or so years transportation system managers have received data detailing system performance from a range of infrastructure-based sensors such as inductive loops or CCTV cameras. While these sensors have enabled performance increases to be garnered from transportation networks, the data they provide is limited in scope, timeliness and availability. However, the explosive growth in mobile devices and communications network capability has created a fertile ground for new developments in transportation-oriented sensing. It is the purpose of this chapter to first take a look back at extant infrastructure-based sensors and then move on to positioning, communications, participatory sensing and finally data management issues. These, as discussed in Chap. 1, form the basis for the first tier of the DMII which underpins the new transportation paradigm outlined in this book.

2.2 Traffic Detection and Surveillance Systems

The ability to have a complete picture of the environment is an essential ingredient for making intelligent operational decisions. In this section, the most commonly deployed infrastructure-based traffic sensors are discussed before moving on to in-vehicle and mobile sensors and concluding with a discussion of some newer traffic sensing technologies.

2.2.1 Sensors for Traffic Monitoring

Traffic detection uses a variety of technologies including acoustic, video, ultrasonic, microwave radar, infrared and other modalities. The earliest instances of automated

P. (Vonu) Thakuriah and D. G. Geers, *Transportation and Information*, SpringerBriefs in Computer Science, DOI: 10.1007/978-1-4614-7129-5_2, © The Author(s) 2013

detection and surveillance in transportation were for the measurement of road traffic, prior to which measurements were made manually. The authors of [22] note that the monitoring of roads via Closed Circuit Television (CCTV) was first implemented in the US in Detroit, Michigan, in 1961. At around the same time the Chicago Area Expressway Surveillance Project pioneered the use of inductive loop detectors [23], and these remain the primary method of vehicle detection on Chicago's highways to this day.

It is convenient to define two classes of sensors: those that are essentially static in location, such as inductive loops, and those that are highly mobile such as probe vehicles or wirelessly connected sensors. The former are commonly associated with traditional ITS applications and will be discussed first. The latter are a fundamental enabling technology for LBS and Cooperative ITS (C-ITS). Of course, mobile sensors may be used in static scenarios and some traditionally fixed sensors (at least in the road transportation sector) such as cameras and radars may also be used as mobile sensors. Further, static sensing systems may be further classified as either in-road or over-road.

Currently the vast majority of traffic data is captured using static and point-detection sensors such as inductive loops. The adherence to point-detection sensors is largely due to the longevity of traffic control systems and infrastructure with the two predominant systems—SCATS [24] and SCOOT [25]—having been is use and development for more than 25 years. Commercial efforts to produce viable loop replacements have met with only limited success. The reliance on such signal control systems has perhaps hampered both the development and deployment of enhanced sensors and the acceptance of newer traffic control algorithms that rely on a more detailed approximation of the traffic state than can be achieved using inductive loop data alone.

2.2.1.1 In-Road Sensors

For many years inductive loops have been taken as the gold standard for traffic detection despite being disliked by road maintainers who see the required saw cuts as areas of potential sheet failure. From the installation and maintenance point of view, any in-road detection system requires traffic lanes be closed, resulting in traffic disruption and consequently higher costs. The major infrastructure-based sensors are:

Inductive Loop Detectors: The most widely deployed vehicle sensor is the inductive loop [26] which consists of a conductive wire loop (rectangular or circular geometries of greater than 1 m perimeter are typical) buried in the road. The buried coil forms part of the inductive component of a tuned electrical circuit. Vehicles passing over the loop cause the inductance to change and this is detected by associated road-side mounted electronic circuitry. Inductive loops are extremely accurate but can suffer from excessive failure (due to loop breakage) in environments where freeze-thaw cycles are common.

Loops are commonly used to count vehicles, detect presence and to infer headway and occupancy. A pair of loops (sometimes referred to as a split loop) may be used to determine vehicle speed by noting the time at which the vehicle passes over each loop. Speed resolution is a function of the data sampling rate of the detection system and falls as vehicle speed increases. Once the speed is known, vehicle length can be estimated, allowing crude vehicle classification.

Magnetometer: During the last decade or so, in-road battery powered two-axis flux-gate magnetometers [26] have become available. These devices are typically placed under the road surface in the middle of a traffic lane. Battery life under normal usage has been estimated at 10 years [27]. These sensors communicate with the roadside equipment using low power wireless, typically in the 900 MHz or 2.4 Ghz range, and so provide for simpler installation than required for inductive loops. Physically, flux-gate magnetometers sense the change (direction and magnitude) in the Earths magnetic field produced by the presence of a nearby ferrous metal object such as a car.

Like loops, magentometers can detect vehicle presence and so can be used for vehicle counting. Similarly, a pair of closely spaced magnetometers may be used for speed measurement and crude vehicle classification.

2.2.1.2 Over-Road Sensors

All over-road systems suffer from the effects of weather, lighting and occlusion to greater or lesser extents. No single system will ever function "perfectly" under all conditions and much research in the data fusion regime has been entered into. The major technologies used for over-road sensing are:

Passive Infrared: Passive infrared (PIR) detectors contain non-imaging sensing elements that respond to the self-generated and reflected component of infrared radiation from an object. All objects with finite temperatures emit radiation in the infrared part of the electromagnetic spectrum and these sensors tend to operate in the 10 μm wavelength range. Electronics in the passive infrared sensing system establish a nominal background temperature and any object that differs in temperature form the background will be detected. In the case of vehicle sensors, the nominal background is the road surface. Multiple sensing elements are used to create multiple detection zones on the road and passive infrared sensors are able to measure vehicle speed and size in the same way as an inductive loop pair. The PIR is typically gantry mounted over the lane that is being observed, and receding or approaching traffic may be monitored [28].

Active Infrared: There are two types of active infrared vehicle detection systems [26]. One uses LEDs, operating at around 850 nm (in the near-infrared) to illuminate the scene and operates in a non-imaging mode with a receiver similar to the PIR described above. The other uses an infrared laser which is scanned across the detection area. The IR receiver builds up a range image of any objects in the detection area. Active IR systems may also be used in a side-fire configuration and can cover many traffic lanes. Some active IR systems have also been certified for speed enforcement in certain jurisdictions.

Acoustic: Passive acoustic vehicle detection systems are effectively microphone arrays that monitor traffic and vehicle sounds [29]. Such sensors can not only detect vehicle presence but can also, perhaps surprisingly, be used for vehicle classification [30].

Ultrasonic: In these systems, ultrasonic sound waves (with frequencies greater than 20 kHz) are projected into the detection area and the returned signal reflections detected. Such systems may be mounted overhead or side-mounted [31, 32]. Similar to radar in many respects, ultrasonic detection systems report vehicle presence from the reflected signal and vehicle speed from the Doppler shift. Doppler shift is the small change in frequency caused when a wave (electromagnetic or sound) is reflected (or emitted) from a moving object. Doppler shift is explicitly linked with vehicle motion and stationary vehicles cannot be detected [33].

Video Systems: Many public transportation operators and roads authorities have installed video surveillance systems over the last thirty or so years. The earliest systems used predominantly commercial-grade monochrome closed circuit television (CCTV) cameras and in some, slow-scan (several seconds per frame) equipment was used (the analog images were transmitted over leased copper lines in the audio range). The cameras would typically be installed at known trouble spots and the feeds displayed on a video wall located in a Transportation Management Center (TMC) and monitored by human operators. With the correct software, video traffic monitoring systems can provide traffic information from presence through to vehicle identification, with the caveat that the data provided may have a measurable error rate. While many of the monochrome CCTV cameras have been replaced by color CCTV cameras and lately by network (or IP) cameras, they are still predominantly viewed by human operators despite significant recent advances in computer vision technology (discussed below).

Clearly, camera and camera-like sensor systems provide a much richer data stream than that available from the simple in-road point sensors described above. However, the automated generation of useful traffic information from video data, such as vehicle count, queue length or speed, is a non-trivial problem that has been actively studied by the computer vision and ITS research communities since at least the late 1970s (see [34–36] and their references). Such systems vary in complexity from inductive loop "replacement" systems to sophisticated vehicle trackers and classifiers. The latter class of systems also lend themselves to the development of automated incident detection systems [37–40].

There are many challenges to address in these systems. Deployed systems must work robustly in real-time under all weather and lighting conditions. When coupled to an actuated traffic light control system, such devices must know when they are not working so the signals can fall back to fixed-time operation in order to prevent potential crashes caused by frustrated (and delayed) drivers of vehicles that have not been detected.

The techniques used to locate and track vehicles in video frames cover the full gamut of those available to the computer vision community. Traditional motion-detection based techniques using interframe differencing [41] or background subtraction [42, 43] have either been used directly to infer vehicle presence or as preprocessing steps prior to applying more advanced methods [44, 45]. Of course, motion-based methods are indiscriminate and essentially respond to pixel intensity change (albeit with various levels of sophistication) and so cannot readily distinguish a shadow on the road from a vehicle. Indeed much effort has gone into solving this problem [46–48].

Using motion-detection to locate regions or "blobs" of interest, however, is a practical means of vastly reducing the in-frame image regions over which more sophisticated and computationally more expensive vehicle detection algorithms must be run. The more sophisticated and robust methods usually rely on having an internal vehicle model, and use pattern matching techniques to distinguish vehicles from non-vehicles and image background.

The pattern matching formalism generally relies on identifying a set of "invariant features" which are common to all vehicles, of which headlights are a good example [49]. The typical process consists of extracting many image patches containing headlights and many not containing headlights. These "positive" and "negative" examples are then input into the training phase of a supervised learning system of which there are numerous examples [50]. The more fashionable currently are Support Vector Machines [51] and the numerous variations on the Boosting algorithm [52]. The output of the learning system is a function such that when an image patch is input, the output will be "true" if it is a headlight and "false" otherwise. Of course, this is a gross oversimplification as no classifier is perfect. There will be "false negatives" and "false positives" and there is no guarantee that the system will "generalize" to images captured using different cameras or under different illumination conditions. Automatically extracting the image patches to input into the classifier is also non-trivial if not performed naively using exhaustive search over the image. Of course, similar techniques may be used to locate pedestrians and other objects of interest in video feeds.

Continuing on with the headlight example, the vehicle model consists of the vehicle headlights and their spacing in world coordinates, which despite its simplicity allows for some distinction between vehicle classes. The term "world coordinates" requires explanation. Video images taken by a single camera are a projection (usually through some optical system comprising lenses and aperture stops) from the three-dimensional world observed by the camera onto the two dimensional image plane (CCD sensor, film, etc.). If the camera parameters (roughly height, pitch, roll and focal length) are known, then the parameters of the projective transform from world coordinates to image coordinates can be computed [53]. The inverse transform will map image coordinates to world coordinates and thus pixel separation in the image can be converted into real world distances. In the single camera case there is some ambiguity in the transform and care must be taken to ensure the correct mapping is used. The use of two cameras in a stereo pair removes the ambiguity in the transform but requires the additional complexity of matching the slightly disparate images from two cameras before being able to compute the parameters of the required projective transform. There is at least one commercial product that uses a stereo camera for road side monitoring of pedestrian crossings [54].

Computer vision also forms the basis for Automatic Number Plate Recognition (ANPR) systems. If used between two points, such systems can be used to measure travel time and conduct trip origin-destination (O-D) studies [55].

Radar: There are two types of radar systems that have been widely deployed for measuring traffic parameters from the curbside: Doppler and Frequency Modulated Continuous Wave (FMCW). Both provide all basic traffic data and they are discussed in turn.

In Doppler radar, super-high frequency radio signals (at frequencies near 10.5 and 24.0 GHz as permitted for traffic detection by the Federal Communications Commission (FCC)) are transmitted into the traffic stream and the reflected signals detected and processed to determine their Doppler shift [56] and hence vehicle speed. Stationary vehicles cannot be detected.

In FMCW radar, the extremely high radio frequency signals are modulated by a triangular waveform so that the frequency of the transmitted signal varies slightly with time (this is the definition of frequency modulation). The signal reflected from the target is mixed with the transmitted signal to produce a beat of much lower frequency. By processing the beat signal it is possible to extract the range and speed of the vehicles illuminated by the radar [33].

Laser-Pulsed: Pulsed laser systems are typically referred to as LIDAR (an acronym for Light Detection and Ranging) systems and they provide the full gamut of basic traffic data. In these systems, near-infrared laser pulses, typically at the kilohertz rate, are used to illuminate vehicles and the reflections detected. Time of flight of the pulses is used to determine range and the time between successive pulse returns is used to determine vehicle speed. With these systems it is possible to build up a profile of the vehicles passing through the detection zone.

Other Technologies: Various other technologies have been researched [57, 58] to measure the same physical properties of vehicles (electric/magnetic properties, reflectance, sound, etc) as the sensors described above. Weigh-in-Motion systems deserve special mention because unlike the sensors described above they are primarily used for road access control and pricing rather than real-time traffic management. Most systems are based on conventional load cell technology but recently optical fiber-based systems have been proposed [59, 60]. An interesting possibility would be the dual use of a fiber sensor as an inductive loop replacement (magnetic field sensor) and simultaneously for weigh-in-motion.

2.2.2 In-Vehicle Sensors

The number of sensors in a car has also rapidly increased, with sensors for the powertrain (which monitors vehicle energy use, drivability and performance, and involves the engine, transmission and onboard diagnostics), chassis (the main control functions of which are monitoring vehicle handling and safety, and involves the steering system, suspension, vehicle braking and stability) and, finally, for the body (which monitors occupant needs including occupant safety, security, comfort and information) [61].

One direction in automotive systems is in the area of Intelligent Vehicle Technology, a concept typically associated with the development of autonomous vehicle functionality for Unmanned Ground Vehicles (UGV). The key attributes of intelligent vehicles include the following: (1) the ability to sense the vehicles own status as well as its environment; (2) the ability to communicate with the environment; and (3) the ability to plan and execute the most appropriate manoeuvers [62]. Applications of intelligent vehicle technologies to the automotive sector are often seen as the next generation of vehicle safety systems. Specifically, for applications within the automotive industry, "Intelligent Vehicle" systems are defined as systems that sense the driving environment and provide information or vehicle control to assist the driver in optimum vehicle operation [63]. Overall, different data about the driving environment can be obtained through any combination of sources such as on-board video cameras, radars, lidars, digital maps navigated by GPS, communication from other vehicles or the highway infrastructure.

Vehicle-mounted radars, typically operating at 77 GHz, are used in adaptive cruise control systems, overtaking assistance systems and rear-end collision warning systems (which may include brake assistance). Many vehicles are now available with parking sensors that detect nearby objects when the vehicle is travelling at low speed. Some vehicle manufacturers extend these systems to semi-automatic (the driver controlling the speed of the vehicle) reverse parallel parking and recently to 90° reverse parking [64].

2.2.3 Some New Sensing Modalities

Additional sensing systems are now available which are either being used currently or hold potential for future use.

Toll Tags: Many vehicles are now equipped with active Radio Frequency Identification (RFID) tags (known as Electronic Toll Collection or ETC tags in the industry), which are primarily used for free-flow tolling applications. Travel time and O-D information between receiver gantries may be easily obtained by matching tag IDs [65]. However, many jurisdictions will only allow partial tag IDs to be released, which introduces a source of noise in the data and reduces the overall accuracy. The use of ETC information opens up a whole range of privacy issues, many of which are discussed in Sect. 2.6 below.

Bluetooth: Bluetooth is a short range (less than 100 m) wireless technology that is most commonly used to connect fixed and mobile devices together. A typical use is to allow hands-free operation of a mobile phone while driving using factory installed or third party devices. Each Bluetooth radio broadcasts a unique identifier known as the media access control (MAC) address.

By using road-side mounted equipment it is possible to detect the MAC addresses of Bluetooth radios in passing vehicles. Matching the received MAC addresses between spatially separated road-side receivers enables travel time and O-D surveys in an identical manner to using ETC tags [66, 67]. It is to be noted that Bluetooth MACs contain no information that can be used to identify the owner or user of the device, and so privacy is less of a concern. Such systems are now available commercially.

Mobile Phones and Mobile Devices: The rapid growth in the adoption of personal connected devices such as mobile phones, Personal Digital Assistants (PDAs) with wireless connection, handheld devices, tablet PCs (such as the iPad) and smartphones has brought about a sea-change in ubiquitous information generation, sharing and delivery. Mobile phones have sensed traffic since the 1990s, with drivers calling by phone to report traffic incidents to emergency management services. In the last decade, the convergence of several technologies has enabled connected mobile devices to become critical sensors for mobility services, particularly on an ongoing, organized basis.

Many current smartphones have sensors for position, sound, video, acceleration, ambient lighting orientation and proximity. Such devices have large memories and significant processing power. Additional functionality is made possible through third party "apps" which are able to access many of the internal systems through well-defined Application Programming Interface (APIs) [68, 69] which is discussed further in Sect. 2.3.3.

Underlying technologies that have positioned connected mobile devices to become an important sensing approach include being location-enabled (either through built-in GPS or through cell tower multilateration), WiFi connectivity, and integrated functionalities such as the ability to take high-resolution photographs of incidents and special events that have the ability to disrupt traffic. Two important benefits of connected mobile devices are their high penetration rate which enables sensing at significant granularity and density, so as to become "pervasive" sensors of the transportation environment on an as-needed basis; and their ability to be connected, by special instrumentation, to secondary sensors, for example, for air pollution or noise-level monitoring [70], and integration with wirelessly-connected wearable body sensors for health monitoring. Some of these applications are discussed in Chap. 3.

In essence, mobile network-connected devices can act as real-time probes in the transportation network and for location-monitoring for social networks and many other applications. For the developers of in-car navigators and on-line mapping systems, this new source of data has enabled a range of applications from traffic congestion warnings to travel-time estimation.

Biometric Sensors: Biometric sensors are body-wearable sensors that collect, store, and share relevant data. Developments in wearable sensors have led to diagnostic as well as monitoring applications, which allow physiological and biochemical sensing, as well as motion sensing [71].

A new generation of wearable biometric sensors measure Electrodermal Activity (also known as skin conductance or galvanic skin response) for monitoring arousal associated with emotion, cognition, and attention. Another type of biometric sensor is mobile Photoplethysmography (PPG) for screening of cardio-vascular pathologies that may find increasing safety, security and travel quality applications in transportation. Applications are in the areas of mobile health and remote health telemonitoring, for example, while driving.

People as Sensors: One of the most active research areas covered by this book is on the use of humans as sensors. Information generated by users is generally called User-Generated Content (UGC) and can primarily occur *proactively* when users generate primary data on events, concepts or activities of interest; or *retroactively*, by analysts who process secondary user-submitted data that is published using social media tools such as Web 2.0 tools, blogs, microblogs and so on. Several different sub-modes can be differentiated within these modes: for example, proactive UGC can result from idea or design competitions, participatory sensing (where users voluntarily report events witnessed or share mobility experiences) or by opportunistic sensing (where users agree to be "tracked" and allow movement, trajectory or speed data to be transmitted to peers or to a central server). We review these modes of people-centric sensors in Sect. 3.3.

2.3 Transportation-Oriented Communications Systems

During the period of explosive growth of in-vehicle sensor technology a similar exponential growth in wireless communications technology has also taken place. The ability to share raw sensor data or derived information with other vehicles and travelers in (or near) real time enables a range of previously unforeseen transportation applications.

It is beyond the scope of this work to go into the details of mobile data communication standards and techniques [72]; however, it is worth noting that travelers can be permanently connected to an Internet Protocol (IP) network with a bandwidth of several megabits per second for less than $10 per month; and this cost will only fall as time goes on. The development of these systems is led by the large telecommunication companies and follows consumer demand.

2.3.1 Communications Access for Land Mobiles

A systems architecture based on IP Version 6 [73]—Communications Access for Land Mobiles (CALM) [74]—is being developed for continuous communications between vehicles (Vehicle-to-Vehicle or V2V) and between vehicles and infrastructure (V2I) which spans a range of data rates and latencies by seamlessly switching wireless mode (physical layer) based on need. For example, messages pertaining to safety should have priority over messages reporting headlight status. Of course, pedestrians can be easily accommodated through connections to nomadic devices such as cell phones. A highly simplified functional diagram of the CALM architecture is shown in Fig. 2.1.

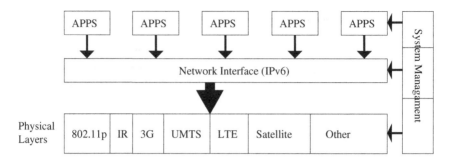

Fig. 2.1 Simplified CALM architecture

As can be seen from Fig. 2.1 CALM can easily accommodate wireless mobile data services such as the General Packet Radio Service (GPRS), Universal Mobile Telecommunications Systems (UMTS) and Long Term Evolution (LTE) over GSM; and the various data transmission additions to CDMA. Satellite communications may also be readily included, as can any IPv6 capable bearer.

The now standardised low latency component of CALM is designated CALM M5 (ISO 21215) which extends IEEE 802.11p (known as WAVE—Wireless Access in Vehicular Environments—in the US) and is colloquially referred to as Dedicated Short Range Communications (DSRC). WAVE is part of the IEEE 802.11 family of wireless standards [75] of which the familiar consumer WiFi (802.11a/b/g/n) is a member. Operating at 5.9 GHz, DSRC offers a communications range of up to 1 km. A complying DSRC implementation based on CALM M5 or 802.11p broadcasts vehicle position and other parameters at 10 Hz in the so-called Cooperative Awareness Message (CAM—colloquially referred to as DSRC Beacon Messages). It will not be long before mobile phones are equipped with DSRC radios and so link pedestrians into the mix.

It is envisaged that vehicles and other travelers will form part of a wireless Mobile Ad Hoc Network (MANET) based on DSRC. The adjective ad hoc is included to indicate that the clearly decentralized network does not rely on any existing infrastructure and members (nodes) may come and go over time. Further, each node forwards messages it receives to their ultimate destination based on a routing algorithm. Mobile routing algorithms are the subject of much research.

When specialised to vehicles, MANETs are given the moniker VANET [76] for Vehicular Ad Hoc Network. Interest in VANETs has been largely motivated by safety and traffic efficiency considerations and the possibility of opening up a whole new area of innovative applications. Research in VANETs has focused on the inherently unfavorable characteristics of the wireless communications environment in which VANETs have to perform, the fair and efficient use of available bandwidth in a totally decentralized and self-organized network, the high mobility and scalability requirements, the wide variety of environmental conditions in which these networks have to perform, and the potential security and privacy considerations of receiver and sender. Research into these areas has significantly increased in the last two decades [77–79].

Several prominent, primarily government initiatives have provided opportunities to researchers to study the technical and socio-economic aspects of VANETs, such as the Connected Vehicles (formerly IntelliDrive^{SM}) program in the US, and four integrated projects within the 6th Programme Framework of the EU in areas that touch the field of VANET: COOPERS [80], CVIS [81], PReVENT [82], and SAFESPOT [83].

2.3.2 Machine-to-Machine (M2M)

With the ever increasing number of network-centric sensors and systems with large computing power, the requirement for human intervention in the sense-communicate-control chain falls, ultimately to zero and the need for direct Machine-to-Machine (M2M) interaction becomes self-evident. According to [84], M2M represents a future where "billions to trillions of everyday objects and the surrounding environment are connected and managed through a range of devices, communication networks, and cloud-based servers".

The initial concept of the "Internet of Things" (IoT) attributed to Ashton (and promulgated by The Auto-ID Labs [85]) which was based on RFID tags barely scratched the surface of what is now possible with the advent of 802.11 and the low power IEEE 802.15.4 ("Zigbee") [86] wireless protocols.

Mobile devices with cameras capable of reading QR codes (somewhat coincidently invented by Denso Wave, a subsidiary of Toyota, for tracking vehicles during manufacture [87]) also provide novel ways of connecting unintelligent objects into the mix. Of course standardization has played, and continues to play, an enormous role in the adoption and deployment of such technologies. In the context of this work it is clear that v2x systems; and traffic and transportation sensor systems fit neatly into the M2M (or extended IoT) paradigm.

Connecting large numbers of sensors together enables much greater knowledge of the system state but brings with it significant challenges especially in device management and control. The limited address space provided by IPv4 is clearly a problem of the past but must be handled when dealing with legacy systems. Sensor nodes which "sleep" for significant periods do not fit into the conventional IP network modalities. Nor does the small packet size prescribed for Zigbee. However, neither of these challenges is insurmountable [88, 89].

The proprietary Urban Operating System (UOSTM) [90] and other middleware systems such as Ice from ZeroC [91] provide M2M integration of services by means of a programming platform and operating infrastructure for the development of sensor networks of intelligent devices embedded in urban systems and distributed sensing applications that run on them.

The Web of Things (WoT) redefines the IoT to use the standard web protocol HTTP and the RESTful architecture [92] for communicating between devices. Not only does this make it easier for programmers to connect to a wide range of devices through well-known means, it also makes it easier for human interaction with potentially unfamiliar devices through familiar technologies such as the Web Browser.

Even then human interaction with the WoT can be challenging. How should the data be presented? What visualization techniques are helpful when the dimensionality runs into the billions? Where, when and how should humans intervene?

2.3.3 Application Programming Interface

The data collected by sensors of all modalities must be made available to programmers and users if it is to be useful in developing new and novel systems. Programmatic interfaces have tended to be formalized through an Application Programming Interface (API). Formally, an API is a source code level interface that enables programmers to connect software components together. APIs may be programming language and operating system dependent or independent even though they usually provide an interface to an object-code library targeted for a specific architecture. In the network-centric world of today the concept of Service Oriented Architectures (SOAs) has arisen. An SOA is a programming paradigm in which a network-based service (usually a web—either html or xml—service) provides specific functionality. Software as a Service (SaaS) is an extreme example of a SOA but the input or output of a (web) service do not have to be human readable. The power of SOA lies in the ability to connect multiple services together in order to obtain functionality beyond that which the component services provide separately. Clearly the component services of a SOA need a well-defined and documented API in order to be useable.

In fact, designing usable APIs is critical for any data organization exposing a programmatic user interface. One of the most important reasons is that highly usable APIs can drive adoption and sustained use of a particular technology and initial encounter with a poorly designed API can discourage users from ever adopting a particular technology. However, many data sources which are useful for mobility services are not yet associated with APIs. Sometimes such data is available through Rich Site Summary [93] (RSS) or its (almost) successor, Atom Syndication [94, 95], feeds.

A number of transportation agencies have released public APIs to access web feeds (see [96] and [97] for example). Judicious use of the reader's favorite search engine will reveal many more.

2.3.4 Positioning Systems

An essential ingredient for many of the above applications is the need for location-awareness through positioning systems. It was noted in 1993 that the first automotive in-vehicle navigation systems introduced in 1984 utilized a combination of dead reckoning and map matching to track a vehicle's movement over the road network [98]. Dead reckoning was accomplished with differential wheel sensors and a magnetic flux-gate compass. Map matching required highly accurate digital road maps and sophisticated software to correlate vehicle motion with the road network and to deal with the many subtle complexities involved. New sensors and technologies including inclinometers, gyroscopes, satellite-based navigation, inverse Loran, roadway electronic benchmarks, electronic odometers and ABS wheel sensors have been considered to supplement or replace the special wheel sensors and magnetic compass, and to simplify or eliminate map matching.

Modern outdoor positioning systems are either satellite or mobile-phone based. The satellite based systems (such as the currently available US Global Positioning System (GPS) and Russian GLONASS systems, and the European Galileo or Chinese BeiDou-2 systems that are under development) are generally more accurate and available anywhere on the surface of the Earth. These satellite systems require line-of-sight visibility from the satellite to the receiver, which may be unattainable in urban areas (the so called "urban canyon" effect) or in dense forests. Under ideal conditions, satellite systems exhibit position errors of the order of ten meters or less. *Satellite-based Positioning*: The US GPS may be treated as the archetypal satellite based positioning system. The other systems named above use the same fundamental positioning technology based on intersecting range-spheres from several satellites. In general, three satellite signals are required for an accurate position fix. In its raw form, GPS has an accuracy of about 10 m horizontally and 20 m vertically, which is sufficient for navigation but not for vehicle collision avoidance.

GPS accuracy may be augmented by broadcasting corrections to GPS positions using terrestrial or satellite based transmitters [99]. The terrestrial system is termed differential GPS (DGPS) and the satellite based system, which is available throughout the continental US and Alaska, is known as the Wide Area Augmentation System (WAAS) [100]. Both systems rely on computing GPS corrections from accurately surveyed reference stations and differ only in the means of disseminating the corrections. WAAS typically enhances GPS positional accuracy to 1.0 m horizontally and 1.5 m vertically over the reception area. Terrestrial DGPS has notionally zero error at the reference site with the error growing linearly with distance from the reference.

A novel technique for GPS correction, called the Quasi-Zenith Satellite System (QZSS) [101], is being developed in Japan and will ultimately use a three satellite constellation to provide sub-meter class positioning. As well as transmitting DGPS data, QZSS will also transmit supplemental GPS signals. The satellite orbits are designed so that at least one satellite will be visible at an elevation of 70° or more above the horizon over the Japanese islands, thereby reducing the effect of urban canyons on positioning accuracy.

Mobile Phone based Positioning: Mobile phone based positioning systems rely on signal multilateration between several mobile phone towers and the handset [102, 103]. Typical accuracy is of the order of fifty meters. Positioning at such poor resolution is unsuitable for navigation but is suitable for other less stringent location-based service requirements. It should also be noted that the majority of smart phones are equipped with in-built GPS receivers and use a combination of multilateration and GPS to provide accurate positions. Additionally, most GPS equipped handsets utilize Assisted GPS (A-GPS) [104] in which the mobile phone service provider transmits various GPS related data to the handset thereby speeding up the time the user takes to obtain a GPS position.

Other Issues: For some cooperative safety applications, however, such as side-impact collision warnings, the accuracy achieved by the above technologies is not sufficient, and research into real-time kinematic (RTK) positioning systems is being conducted for improved relative positioning at centimeter-level accuracy [105]. Similar accuracy may be possible by broadcasting differential corrections from DSRC road-side units or on sub-carriers of commercial radio or television signals. Positioning is also possible through technologies such as WiFi and Bluetooth, although they have been used to a lesser degree in transportation.

Indoor Positioning: One area of recent research interest is the area of indoor positioning, a problem that is motivated by GPS signal attenuation inside buildings. These types of applications are relevant to transportation particularly for mobility planning through mobile devices inside rail and transit terminals; and airports, for pedestrians looking to walk through public buildings to shorten their travel time, and for E911 emergency planning. Augmented Reality (AR) technologies have been used to supplement location-awareness in indoor positioning. AR is a powerful user interface technology that augments the users environment with computer generated entities and is defined by three important aspects [106]. It blends the real and virtual within a real environment, is real-time interactive and registered in 3D. In contrast to virtual reality, which completely replaces the real world, augmented reality displays virtual objects and information registered to real world locations. These technologies have significant potential for persons with visual impairments and other types of disabilities.

2.4 Methods to Add Intelligence to Sensor Data

Virtually all sensing technologies require processes to extract intelligence from the raw data, which may not be useful for transportation applications by themselves.

Information Extraction Methods: Methods that translate raw output from sensors are as numerous as the sensors themselves, and many of the mature technologies have been extensively studied in other disciplines. Nevertheless, what works and what does not work with many of these mature detection technologies have undergone significant study and testing by the transportation community, given considerations of purchase, maintenance and repair costs, the types of traffic parameters that can be monitored (for example, some types of sensors such as inductive loop detectors can measure traffic volumes, speed with the use of two loops, vehicle type, occupancy, and headway between vehicles, but not the density of vehicles over a stretch of highway, whereas others, like video cameras, can measure all these parameters), the need to be obtrusive in some cases and unobtrusive in others and operational performance under different weather conditions (for example, see Weil et al. [22], for an evaluation of sensors for the purposes of automatically detecting traffic incidents).

Information extraction methods can result in a range of intermediate measures, such as simple measures of counts (of persons, cars, pedestrians during some time interval) to more complex frequencies (number of occurrences of specific words, number of interactions among persons during a day, or trips to a certain destination between a certain period of time), continuous measures such as speeds (including time mean speed or averages of speeds of cars or other moving objects crossing a point, or space-mean speed, which the average of the velocities by which moving objects crossed a segment of a road), trajectory, or weight (for weigh-in-motion technologies primarily used for trucks), as well as classified measures into nominal categories (automatic detection of a car versus a truck; or identification of a GPS sensor in a slow-moving bus versus a pedestrian) or ordinal categories (such as different intensities of traffic congestion such as heavy congestion or stopped traffic or precipitation levels such as thunderstorms versus light rain).

An example of an information extraction processes that has received considerable attention in the transportation literature is Video Image Processing (VIP) which converts raw video streams into traffic data. Automatic recognition of speech has been used for many years in phone-based traveler information systems, for controlling in-vehicle functions, for sending text messages or name dialing through mobile phones, and by operators to respond to calls about public emergencies.

More recently, web-based and social media information sources have led to the generation of massive amounts of structured and unstructured (text, video, still images) "Big Data" that are currently active areas of research. Transportation, as does any other sector, generates a large amount of material in unstructured free-form text, such as plans, regulations, policies, technical documentation, and media and user-generated web content. Content analysis has been used on text material for qualitative assessment text-based content, but the use of text mining, using multi-disciplinary aspects of information retrieval, text analysis, information extraction, clustering, categorization, visualization, database technology, machine learning, and data mining, has been used to a lesser degree, although there are some examples of using text mining to develop ontologies of urban planning documents [107] and of microblogs for real-time event monitoring [108].

One development stimulated by the generation of mobility data such as GPS trajectories and advances in data mining technologies is mobility mining whereby digital trajectories are processed to understand people's mobility and activity patterns, for purposes such as mobile commerce, location and context-based search and advertising, early warning systems, traffic planning and management and route prediction.

Analysis: In most cases, raw information from sensors has little meaningful end-use value and needs to be processed in various ways for different purposes. Mobility analytics uses sensor data using a range of methods to utilize data generated by the underlying sensor and communication technologies. Virtually every aspect covered in this book is associated with information processing and analysis which draw from methods in transportation engineering and planning, operations research, management, computer science, geographic information science, and several social sciences and related disciplines, to process the raw data and detect and understand patterns in mobility data and to build services, plans and management strategies, using the intelligence derived.

Information Fusion: Information fusion, of which the perhaps more familiar Sensor or Data Fusion are subsets, is a well-researched topic concerning the exploitation of data from multiple sources including sensors, databases, user generated content, etc. in order to obtain a better view of the world than would be possible from any single source alone. In a work such as this, it is barely possible to scratch the surface. It should be noted data fusion is termed data integration in the GIS domain.

In essence sensor or data fusion is a principled way of combining data from multiple sensors to yield better results than any single sensor could produce on its own [109]. A simple example is afforded by considering an in-road inductive loop and a calibrated road-side video based vehicle tracking system whose field of view includes the loop. Both sensors are able to provide estimates of vehicle speed within some error bounds. Combining the speeds using classical Bayesian data fusion techniques will improve both the accuracy of the measured speed and its variance [110, 111]. In general, a sensor model will include an error distribution which is essentially a formalization of how good the sensor is. When formulated as a probability density function, this becomes the prior probability in the Bayesian framework. More advanced fusion techniques such as Kalman or particle Filters can also be used to fuse multiple sensor inputs [112, 113].

Event Stream Processing and Complex Event Processing: Event stream processing and complex event processing [114, 115] are two relatively new research areas that are potentially very useful in the transportation regime. An Event Stream is defined to be a linearly (usually by time) ordered sequence of events, say bus arrival times at a particular stop. It would be possible, for example, to predict the arrival time of buses at the stop from the past arrival times or to raise a red flag when two buses arrived less than five minutes apart. An Event Cloud is a partially ordered set of events (which may be unbounded), where the partial orderings are imposed by the causal, timing and other relations between the events. In other words, an Event Cloud can consist of many Event Streams. A simple example of an Event Cloud would be the train arrival times at a transportation interchange and the bus departure times form the same interchange. A more complex example would be the Event Cloud consisting of all available data from the transportation network of a city.

Complex Event Processing (CEP) sits inside the discipline of data fusion but includes methods such as rule-based inference and adaptive neural networks which are usually seen as part of Artificial Intelligence (AI) or machine learning. The essential aim of CEP is to look for patterns that correspond to events that are significant. An example would be the inference of a burst water main, with high probability, if there is a continuously "on" loop detector present at an intersection *and* a bus fails to arrive at a stop nearby *and* there is a drop in water pressure in a main pipe. Determining that a collection of simple events forms a reportable super event is a significant research challenge. Information fused from multiple sources may form one or more of the event streams in a CEP system. Ligozat, Vetulani and Osinski [116] propose a methodology for dealing with CEP in a spatio-temporal setting. Despite being concerned with monitoring for security purposes the same processes may be extended to the transportation domain with appropriate modifications.

2.5 Initiatives and Programs Through Technology Integration

There are several initiatives and programs through which the use of information technology in transportation can be organized. As discussed in Chap. 1, there is no stark line that divides the technology initiatives discussed in this section into mutually exclusive categories. They all utilize information technology, sensor data and communications technology but differ in the way they are organized and managed (public versus private) and the extent to which they emphasize aspects of government-owned and managed infrastructure versus privately initiated applications utilizing advancements in technology

Cross-cutting these technology initiatives are different theoretical and research initiatives that examine various aspects in the transformation of transportation systems, human behaviors and society, as information itself, or information-based solutions become available.

2.5.1 What Happens to All that Data?

The sensors and sensor systems described above will clearly generate huge amounts of data that have both spatial and temporal content as emphasized in the concept of Tier 1 of the DMII. When used in transportation, such databases must function on two different time-scales that embody the movement of travelers throughout a geographic region using the various transportation modes and the slow changes to the transportation network itself. These two time-scales would correspond, for example, to the GPS position of a traveler and the map updates provided by a GPS vendor respectively.

Regular databases cannot handle such data efficiently and a whole class of specialised spatio-temporal databases that utilise special-purpose spatial and temporal data models have arisen [117, 118]. Most commercial [119–121] and many Open-Source databases [122, 123] support spatiotemporal data either natively or through add-ons. Such systems are standardized through the Open Geospatial Consortium [124]. These systems all support spatial queries and predicates (Is there a restaurant within 1 mile of (x, y)?), distance functions, etc.

From the design point of view, spatio-temporal databases generally consist of a spatial database combined with a temporal model that allows temporal integrity of the data to be maintained and enables temporal queries. The most commonly used spatial data structures are based on the R-Tree [125], which represents spatial data as a tree-structured hierarchy of minimum bounding rectangles (the idea generalizes to higher dimensions by using minimum bounding boxes). There are some temporal features in the latest Structured Query Language (SQL) standard (ISO/IEC 9075:2011) but they are far from being routinely available.

It should be noted that most transportation data for transportation operations, traffic measurements and vehicular applications, are useful for only a finite period and that internal spatiotemporal database operations would expunge (or at least move to a backup device) expired data for reasons of storage and search efficiency. The data that has expired for real-time applications, could of course be useful for looking for transportation trends, patterns and associations, and as inputs to the planning process, together with information from Tiers II and III of the DMII, for example. These possibilities are examined in Chap. 4 (Sect. 4.4.2).

2.6 Privacy, Trust and Security

Privacy, trust and security are concepts that are essential to human societal inter-actions. In the case of mobility services, the private information possessed by a traveler potentially comprises their identity, current location, origin and destination of travel, journey time, locational preferences and so on. LBS for transportation usu-ally require a traveler to surrender their (potentially extremely) accurate spatial and temporal location to a transportation service provider in order to obtain a useful ser-vice. However, the potential for misuse of this data exists. In the now classic paper, *Geoslavery* [126], Dalton and Fisher discuss numerous situations where knowledge of position is real power and it is imperative for designers and providers of LBS, with the help of government legislation, to curtail such misuse. Such data should only be sent to a trusted organization via a secure communications channel where it will be stored securely.

Privacy is a fundamental human right. However, privacy is also not a static, immutable constant. People are likely to trade off some privacy protection in return for utility gained, security benefits or risks minimized. Such trade-off notions have been explored in various contexts (for example, [127] in the case of data publishing and [128] in the case of rail travel security). Threats to locational privacy include the risks of unauthorized access to raw location data, location information about a person that is secondarily derived or computed using the raw data, hijacking the loca-tion transmission channel, and identification of the person who is using generating location information while using a web service or mobile device.

There are three fundamental approaches to addressing locational privacy: legal, consumer awareness and technology-based. Legal and policy-based strategies and privacy principles (ability to opt out, user consent, data protection) as well as issues of consumer education and awareness are described in detail in Sect. 4.2.2 Technology-based approaches are numerous; one particular methodology that has gained consid-erable support to address location privacy threats is the idea of privacy-by-design to ensure that collected data is only accessed for the purposes which the user agreed. The idea is simple: privacy concepts must be built into physical and software systems, and business processes from the ground up using Privacy Enhancing Technologies

(PETs). Application of this simple idea would help to reduce the number of privacy breaches that are occurring more-and-more frequently.

While a number of PETs for locational privacy have been proposed ([129, 130] are examples from a voluminous literature), commonly used approaches utilize pseudonyms for anonymity (anonymizing proxy for all communication between users and applications) and location protection by degrading the accuracy of the user's location using geographical and temporal masking or cloaking, and encryption. These approaches, while guaranteeing to a certain degree that precise location information transmitted by a user cannot be easily used to re-identify the subject, also have numerous issues; for example, de-identification in a system using pseudonyms is possible by correlating where any given pseudonym spends most of its time and who spends more time than anyone else at any given location [129], which has implications for frequency of pseudonym update. PETs for mobility application are likely to integrate such privacy-enhancing techniques. For example, in the mix-zone approach, geographical regions are considered where no provider can trace user movements. Users, upon entering a mix-zone, receive a pseudonym that changes when they exit the zone. Users also exit the mix-zone in an order different from their order of arrival. In this way, the identities of users entering the mix-zone at the same time are mixed in a way such that the mapping between their old and new pseudonyms is not revealed.

Other approaches utilize communication aspects to address locational privacy. For example, the Mist routing protocol [131] for mobile users address the problem of routing messages to a user's location while keeping the location private from the routers and the sender by utilizing a set of "mist routers" organized in a hierarchical structure that effectively creates a "mist" to conceal from the system and other users by using pseudonym-based routing. The Onion Routing protocol [132] encrypts messages with multiple keys to form an "onion" around the message. The result is that Onion routers remove a layer of encryption to uncover routing instructions to send to the next router where this is repeated, thereby sending the message through several network nodes. This approach can deter the discovery of both the source and the destination information of the packet.

It is also possible to address locational privacy directly in position sensing systems. For example, the Cricket system [133] is an RFID and ultrasound-based indoor positioning system where location sensors are placed in the user's mobile device so that users listen for their position thereby avoiding disclosure of their location during the location-determination process and without having the need for a centralized location-determining system.

Vehicular networks are designed to be open to all participants. In many cases the users of such systems will be unaware (and may not care) of what data are being transmitted and to where. The 802.11p standard specifies that broadcast messages must not be traceable to a specific on-board unit, that in-vehicle units change IP address when they move to another road-side unit, and that MAC addresses be randomly generated from a local address space. Messages are also encrypted using a public key infrastructure with the certificates being distributed by a trusted authority (more than likely a Department of Transport (DoT) or some other government agency).

The purpose of all these measures is to create positional anonymity and prevent tracking. But is this really achievable? A DoT recording DSRC CAM transmissions would be able to track a vehicle based on some simple traffic flow principles even through MAC and IP address changes, and an anonymised vehicle whose origin and destination are repeatedly (perhaps daily) the same geographic coordinates is really not anonymous.

Being wireless and IP based, VANETS are subject to all the usual network security threats: address spoofing, man-in-the-middle attacks, and so on. In the case of safety related issues, signal jamming (even accidental due to arc-welding equipment for example) must be seriously considered if drivers begin to rely heavily on the system for their personal safety. Sending fake messages from a vehicle could create traffic chaos. Authentication cannot prevent a stolen vehicle being used to generate fake messages. Certainly, information fusion and complex event processing will help to rule out some fake messages and to add weight to genuine messages, but these techniques will not work in all cases. Recent work on privacy in vehicular networks are given in [134, 135].

Ultimately, such systems are built on trust. If the user trusts the organization that is receiving her personal data and is not using it to violate privacy and other concerns, then she will continue to use the system. Issues of trust management are explored further in Sect. 4.2.5.

Chapter 3
Technology Systems for Transportation System Management and Personal Use

3.1 Introduction

In this chapter, we discuss systems and services built using the sensor and communication technologies discussed in Chap. 2. In Sect. 3.2 we describe transportation system management technologies including those targeting congestion management, safety, structural health monitoring, dynamic resource management, and systems connecting transportation to weather management, emergency and crisis management, energy and smart cities. In Sect. 3.3 we discuss people-centric systems where citizens participate in data collection and in generating content for mobility services. This is followed by a discussion of mobility and travel services in Sect. 3.4 including traveler information and location-aware services, mobile health technologies and assistive technologies for persons with special needs.

3.2 Transportation System Management, Operations and Safety

The potential of ICT to manage transportation infrastructure and travel demand are perhaps two of the biggest motivating factors behind ITS. ITS were originally offered as solutions for the management of traffic congestion, by using advanced technologies to manage the capacity of existing transportation systems efficiently. Several aspects of the ITS program have focused on congestion management. One example is adaptive traffic signal control systems (in Sect. 3.2.1.1) where the traffic signal timings at road intersections are updated based on real-time vehicle arrival patterns. Another example is dynamic ramp metering (Sect. 3.2.1.2) where traffic volumes entering into the freeway are effectively metered based on real-time information on the mainline freeway. Yet another example among many other approaches to effectively monitor traffic conditions and make decisions that benefit system use is

P. (Vonu) Thakuriah and D. G. Geers, *Transportation and Information*, SpringerBriefs in Computer Science, DOI: 10.1007/978-1-4614-7129-5_3, © The Author(s) 2013

automated road incident management systems (Sect. 3.2.1.3) where road incidents (such as vehicle crashes or stalled vehicles which block a lane) are automatically detected and their duration estimated.

3.2.1 Transportation System Management

Traffic signals are perhaps the most widely deployed system for alleviating congestion on arterial (or surface) roads. The first manually operated signals were deployed in London, outside the Palace of Westminster, in 1868 and the first fully computerized system was operational in Toronto in 1963 [136]. Despite ever increasing computing power and traffic sensing capabilities, the vast majority of deployed signals run under some fixed time plan, which may or may not be optimized based on traffic flow data.

It is convenient to introduce the concept of a signal phase in which several non-conflicting movements at a road intersection are grouped together and controlled by traffic signals. With reference to Fig. 3.1, which depicts a simple intersection, Phase A and Phase B may be defined as shown. In the English speaking world, the vast majority of traffic signals are run in a cyclic pattern that repeats over extended periods of time. The amount of green time apportioned to each phase defines the phase split and may be expressed in absolute or relative terms. The period of the signals is then simply the sum of the time spent in each phase. Each phase is typically assigned a minimum and maximum green time for safety and efficiency, respectively.

A signal group refers to a set of signals that can operate independently of any other. In Fig. 3.1 there are twelve signal groups if the red and yellow lanterns are counted as individual signal groups. Typically only groups that contain a green need to be enumerated reducing the number of groups to four, one for each approach. Defining signal groups also allows an alternative definition of signal phase: A signal phase is a collection of non-conflicting signal groups; that is, the signal groups that comprise a phase may all show green at the same time while maintaining safety at the intersection. In many jurisdictions, cross traffic turns ("filter" turns) are allowed even though these violate the safety criteria. In fact traffic signal setup comprises a mixture of both (physical safety) law and (local traffic) lore in most jurisdictions.

The standard signal roundels may be extended by arrows that allow or disallow turning movements and also various combinations of signals that control public transportation modalities such as trams and buses. When adjacent or nearby intersections are signalized it may be appropriate to coordinate signal operation to achieve better traffic management outcomes. Coordination requires the definition of offset which is the start delay ascribed to a downstream intersection in order that vehicles leaving the upstream intersection will not be stopped by a red signal at the following intersection. A well-known example is the so-called "green-wave", in which a platoon of vehicles is marshaled through a sequence of intersections that all turn green at the "right time".

Traffic signal control systems may be divided into two broad classes: Phase Control (known as Stage Control in the UK) and Signal Group Control (known as

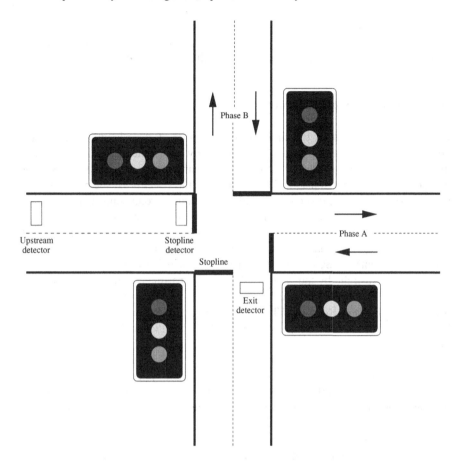

Fig. 3.1 A simple intersection

Phase Control in the UK). In the former, signal phases are switched in a predefined sequence (although some phases may receive zero time) with a fixed cycle time and in the latter each signal group is switched independently and in an acyclic manner. Signal control strategies may be further subdivided into three categories:

Fixed Time; Phase Control Only: Each phase receives the same amount of green on every cycle. No account is taken of the current traffic state at the intersection. The cycle time and phase split may be set by experience or from data collected from a traffic survey. Software such as TRANSYT [137] may be used to determine the optimal timings.

Actuated Control: In these systems, the signals run green on one phase unless vehicles are detected on a conflicting approach whereupon the signals will switch in order to allow the waiting vehicles to proceed (this may not occur straightaway because of time constraints on the signal operation)

Adaptive Control: This strategy is discussed below. Actuated control is fully subsumed by adaptive control. Deployed systems also tend to have public transportation and emergency vehicle priority capability as standard features.

3.2.1.1 Adaptive Signal Control

In adaptive signal control, the choice of one or more of cycle length, split, offset, phase to run or signal group to run is determined by a computer algorithm that bases its decisions on measurement of the traffic state using one or more traffic sensors of the types described previously in Chap. 2. Most of the systems use the phase control methodology; however, it should be evident that signal group control allows better utilization of the intersection but may require additional driver awareness.

The two most widely deployed adaptive traffic signal control systems [138] (although the data is primarily for the US it may be taken as representative of worldwide deployment) are SCATS (the Sydney Coordinated Adaptive Traffic System) developed in Australia by the New South Wales Department of Motor Transport in the 1970s [24] and further refined by its successors (the Department of Main Roads; the New South Wales Roads and Traffic Authority and currently, New South Wales Roads and Maritime Services [139]) and SCOOT (Split Cycle Offset Optimisation Technique) developed by the United Kingdom's Transport and Roads Research Laboratory also in the 1970s [25] and further refined by its successors (the Transport Research Laboratory and now TRL Ltd. [140]). Each system controls more than 35,000 intersections worldwide. Both are traditional signal phase controllers and rely on a central computer system to perform the majority of operational calculations. In addition, both are proprietary systems and their detailed operations are trade secrets. However, it is known that SCATS uses a heuristically based control strategy, whereas SCOOT has an in-built traffic model similar to that used in TRANSYT [141].

The systems also differ in sensor (usually inductive loops) requirements: SCATS requires stopline loops and SCOOT upstream loops. It is evident from the loop requirements that SCATS can only be aware of the number of vehicles leaving an intersection after the signal has turned green, whereas SCOOT is aware of vehicles approaching the intersection. Bearing this in mind, SCATS may be classified as a reactive system and SCOOT as a proactive system. Both systems dynamically adjust cycle lengths and splits; however, SCATS uses offset plans when coordinating between intersections while SCOOT computes the optimal offset based on traffic conditions. Despite the differences, recent work has shown that the two systems produce virtually identical outcomes in enhancing traffic efficiency [142].

Around the same time as SCATS and SCOOT were developed, the Swedish National Roads Administration began development of the LHOVRA system [143] (effectively translated as transportation and platoon priority, main road priority, dilemma zone sensing and red extension system) for isolated intersection control. While not adaptive (in the true sense of the word) this system is a true signal group controller with the main focus on safety rather than traffic efficiency. Additional detectors are placed in the dilemma zone (at a specified speed the dilemma zone extends upstream from the stop line to the point where a vehicle travelling at that speed will not be able to stop if the signal changed to red) and the yellow time extended if a vehicle is detected. Other innovative functions include the ability to change the signals in a green-yellow-green sequence. One disadvantage of LHOVRA is a slightly longer delay experienced by traffic entering or crossing from the minor road.

The rapid growth in computing power and sensing technologies since the 1970s has led to the development of a large number of adaptive traffic control systems with much greater computational requirements than SCATS or SCOOT. Such systems tend to fall into the proactive category and rely on modeling the traffic to varying degrees of complexity. Representative systems that have been deployed include RHODES [144], UTOPIA [145] and BALANCE [146]. The US Federal Highway Administration (FHWA) had sponsored the development of a Real-Time Traffic Adaptive Control System (RT-TRACS), which is no longer an active program, but which, based on actual or predicted traffic conditions, would have been able to automatically choose the appropriate traffic signal control strategy from a suite of control schemes and to monitor their performance, including previously or independently developed methods such as Real Time Hierarchical Optimized Distributed Effective System (RHODES) [147], and Optimized Policies for Adaptive Control (OPAC) [148]. Interestingly, ACS lite [149, 150], the development of which was also funded by FHWA, is not model based but does rely on sensors at the stopline, upstream and mid-block.

Recent work [151, 152] has looked at the possibility of enhancing signal control using V2I communications. It is then possible to use objectives such as the minimization of travel time through the network. Simulation has shown performance gains over other techniques [153].

3.2.1.2 Ramp Metering

Ramp metering is the application of traffic signals to the entry ramps of motorways in order to smooth the flow on the mainline. A typical layout including detectors is shown in Fig. 3.2. The aim is to apportion green time to the vehicles queued on the ramp so as to keep the flow on the mainline near optimal by controlling the vehicle entry rate. In other words, delay on the ramps allows for better flow on the motorway. Fixed time control will not be discussed.

The two most prevalent traffic responsive ramp metering strategies that have been deployed are the demand-capacity algorithm [154] and ALINEA (Asservissement LINaire d'Entre Autoroutire) [155]. The former is effectively a feed-forward control system that commences operation when the main line occupancy, measured at the merge (downstream) detector in Fig. 3.2, exceeds a preset threshold; whereas the latter is a negative feedback system that uses the difference of the current merge occupancy and the set-point occupancy (usually the critical occupancy) combined with the previous ramp flow rate (measured using the entry detector) to set the ramp flow rate to be employed in the next cycle. Importantly, ALINEA has been proved Lyapunov stable under a wide range of conditions [156].

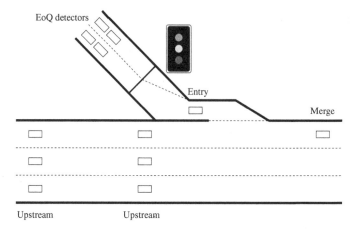

Fig. 3.2 A freeway entry ramp

From a practical point of view, the end-of-queue (EOQ) detectors are necessary to prevent the ramp queue spilling back onto the surface roads and all deployed systems contain methodologies for minimizing this effect with concomitant performance degradation of the metering strategy.

Both methodologies can be enhanced by coordinating ramps along the motorway. There are a number of deployed systems: BOTTLENECK [157], ZONE [158], SWARM [159], SDRMS [159] and HELPER [160] in the US; and METALINE [161], which may be considered as the multi-ramp extension of ALINEA, in Amsterdam and Paris for example. A current notable example is ALINEA/HERO [162] which is a heuristic coordination strategy deployed on 64 ramps over a 75 km stretch of the Monash Freeway in Melbourne, Australia. It was noted that significant performance improvements [163] were gained on a short stretch of the freeway by deploying coordinated ramp metering including improvements in traffic flow of up to 46 % during the AM peak period and 29 % during the PM peak and large improvements of up to 51 % in travel speed during the PM peak period. The average number of crashes per month were noted to have roughly halved after the introduction of ramp metering.

In order to increase efficiency beyond the gains achieved through responsive ramp metering, it is necessary to take into account the non-linear aspects of traffic flow by using some form of traffic model. This should not come as a surprise: the same conclusion was reached for urban traffic signal control systems. The extent to which research on ramp metering is still required is evidenced by the recently concluded EURAMP Project [164]. Many of the proposed systems differ in the choice of traffic model [165] although some learn the model implicitly using machine learning techniques [166, 167]. Very few if any of these methods have actually been deployed in practice.

Enhanced control is possible by including speed control of the vehicles on the mainline [168–170]. Of course greater control can be achieved when the vehicles are fully automated [171]. One area of particular interest that deserves continued attention is the operational integration of freeway and surface road control systems into a unified network control system [172–175]. Such a system would be able to limit spill-back of off-ramp queues without disturbing arterial coordination and use on-ramp metering systems in conjunction with arterial signals to provide traffic load balancing across the network.

3.2.1.3 Incident Management

Sensors on freeways and arterials are frequently used to aid in the detection and verification of traffic incidents toward incident management. It is evident that traffic incidents temporarily reduce the capacity of the surrounding road network and are a major cause of non-recurring traffic congestion. Speed of clearance is of paramount importance for several reasons, not least because secondary incidents due to congestion from a prior crash has been reported to amount to some 20 % of all traffic crashes and account for 60 % of total congestion [176].

The automatic detection of road incidents is an active area of research. The process of incident management is noted to consist of four sequential steps: incident detection, response, clearance and system recovery [177]. Automated Incident Detection (AID) algorithms use traffic sensor data for the detection of road incidents. The voluminous literature on AID algorithms contain examples from pattern recognition, statistical approaches, machine learning, and statistical learning models to detect and decipher unusual patterns of traffic from normally-occurring traffic. Various theoretical AID models have also been proposed; an early example was in the use of catastrophe theory to model sharp changes in traffic flow [178]. These AID algorithms make use of traffic flow measurement made at one point in the roadway segment (typically using loop detectors—thereby being generally referred to as point-based algorithms), in contrast to spatial measurement-based algorithms which make use of video surveillance and VIP [179].

A related problem is incident duration and delay prediction; this line of work focuses on statistical models of incident duration based on measured roadway clearance time; for operational purposes, decision and prediction trees for incident clearance have been developed. These could presumably be qualified by the type of incident, time-of-day, weather conditions and other factors. Sample sizes for these different conditions are likely to be small and hence problematic for robust operations [180]. Most incident management strategies are plan-based and orchestrated through a Transportation Management (or Incident Response) Center (TMC) with the aim being to clear the incident and restore network capacity as quickly as possible. Technical enhancements have revolved around the development of decision support systems that enable TMC operators to make decisions rapidly and accurately, for example, by incorporating traffic simulation capabilities that allow operators to assess the impact of the incident control measures they impose [181] and by reducing the cognitive load on TMC operators by developing multimodal user interfaces [182]. Event mining of User Generated Content (e.g., Twitter, social media) are additional resources for detection of transportation incidents. Examples are given in Sect. 3.3.2.

Work zones should be scheduled well ahead of time and managed much like incidents. Alternative routes must be available to carry the traffic predicted during the lane or road closure, and works should be scheduled outside of peak periods if possible. Unplanned works (such as road failure due to a burst water main) must be treated and managed like a road incident with clearance time and safety being the critical parameters.

In many jurisdictions trucks collide with tunnels and bridges on a more or less regular basis, causing structural damage and induced road congestion due to the frequently long clearance times. Over-height detection is usually carried out by simple break beam systems. Classical means of preventing tunnel ingress include flashing warning signs and traffic signals. DSRC based warning systems that send messages directly to the over-height vehicle have been demonstrated [196].

3.2.1.4 Parking Management

The vast majority of non-commercial vehicles spend most of their time parked. Parking lots consume large amounts of real-estate and in many cases on-street parking reduces road capacity. Motorists searching for parking waste considerable fuel and time, and can increase congestion around shopping and business precincts. Clearly, park-and-ride systems can reduce parking pressure in cities but such schemes need to be incentivized by suitable planning and the availability and reliability of public transport. A single train carriage can account for well over one hundred vehicles given the observation that most cars are single occupant. Pricing parking at an appropriate level is also a vital planning tool.

Many cities provide variable message signs that indicate the number of available spaces in car parks. This is especially common in Europe. Various in-garage systems (for example [183]) have been developed to help guide motorists to free parking spaces and to aid in enforcement and general parking lot management.

The advent of mobile communications systems have made the possibility of en-route guidance to parking spaces possible and DSRC-based systems have been proposed [184–186].

3.2.1.5 Maintenance and Monitoring: Asset Monitoring and Workzone Management

Monitoring and maintaining road infrastructure is a task in which every highway agency is engaged. Failure of major infrastructure can have dire consequences [187] at all levels. Over the last 20 or so years, techniques [188] for continuously monitoring structures in situ have been developed and practical systems are being deployed on an ever greater number of structures. These systems aim to identify potential failures and invoke a proactive response from relevant authorities. Many of the systems use machine learning techniques to determine whether the measured signals differ from the norm [189]. Signal-to-noise ratio is a perennial problem [190].

Highway pavements may have surface or functional failure, where there are cracks and other signs of deterioration and distress on the surface. Failure can also be structural, in that there is a breakdown in one or more of the pavement's structural components or its underlying subgrade to the extent that there is loss of load carrying capacity. There are well-developed procedures in place to regularly inspect and test asset conditions. But given the scale of the pavement infrastructure, these visual inspection approaches and manual testing methods—Destructive Testing (where a section or layer of an asset may be physically removed) or Non-Destructive Testing (NDT) approaches using sensors and other non-intrusive methods—to determine condition can be expensive and inefficient.

Bridge failures may occur due to various reasons including fatigue and brittle fracture of structural members and connections, decay from cracking or corrosion, scouring at the bases of bridge piers where they interface with the stream bed and other factors. From the transportation point of view, the structures most vulnerable to catastrophic failure are bridges, and SHM systems have been deployed on several major bridges throughout the world [191]. Mobile data systems [192] and wireless sensor network technologies have certainly contributed to the increase in the number of such installations due to the relative ease of deployment [193].

SHM allows continuous and autonomous monitoring of the structural integrity and physical condition of a structure using embedded or attached sensors with minimum manual intervention, using non-destructive techniques.

A comprehensive SHM system has the goals of determining changes in critical structural parameters from a baseline, assessing structural integrity and recommending maintenance strategies. Successful implementation of SHM can replace schedule-based inspection and maintenance of structures by condition-based maintenance, thereby reducing lifecycle costs significantly, and improving safety. SHM strategies can be broadly deconstructed into three fundamental yet interrelated and iterative tasks: first, sensor and sensor network development, progressing to fault and damage diagnostics capabilities, and ultimately evolving into prognostication for remaining useful life (RUL) prediction [194].

All SHM systems require a combination of data acquisition from sensors and adequate computational models of the structure. A vast range of specialized sensors are used for these purposes (see [190] for a review). Much of the recent activity has been in developing wireless sensor networks to produce monitoring systems using sensors placed at various locations on the structure. Advances in wireless technology and embedded processing have made wireless smart sensor networks attractive alternatives to wired data acquisition systems on the basis of cost [195]. These technologies have the likelihood to make the civil infrastructures in the transportation system intelligent on a system-wide scale.

3.2.2 Safety

Safety has historically been a primary motivator of transportation policy. Occupant safety is of paramount importance to the automotive industry and is both a stimulator for research and a consumer purchasing incentive. It is the likely predominant market force for the deployment of V2V systems in the not-to-distant future.

During the 1940s it was realized that a large number of frontal automobile crashes resulted in the occupants suffering severe head trauma due to striking the dash or being ejected through the front windshield. This observation led to the development and now (almost) universal deployment of, the three-point seat belt by Volvo in the 1950s [197]. Airbags followed in the 1970s [198], initially as an independent system but now as an integral part of a unified occupant safety system, which also includes multiple structural crumple zones designed to help dissipate vehicular kinetic energy after collision. Automatic raising of head-rests in order to help reduce the incidence of whiplash is a recent addition in some vehicles [199].

While seatbelts, airbags, crumple zones and other in-vehicle systems have certainly saved a large number of lives and reduced the severity of crash-related trauma they fail to address the issue of limiting the likelihood of a collision in the first place. The work of Gazis, Herman and Maradudin [200] on dilemma zones at signalized intersections, which posits a 3-s amber duration at urban intersections and an 8-s amber at rural intersections, may be regarded as ground breaking.

Starting from the mid-1990s, extensive research effort has been expended on looking at ways of preventing collisions from occurring. These may be classified as infrastructure-based or vehicle-based and are discussed in Sects. 3.2.2.1 and 3.2.2.2, respectively. Additional approaches are discussed below including passive and active safety systems, Vehicle-to-Vehicle (V2V) and Vehicle-to-Infrastructure (V2I) approaches; and systems for detection and avoidance of vehicle crashes with cyclists and pedestrians.

3.2.2.1 Infrastructure-Based Safety Approaches

Infrastructure-based systems for preventing collisions have not been widely deployed. While it could be argued that over-height vehicle warning systems and signalized intersections are traffic safety systems, most observers would regard them as infrastructure protection and traffic efficiency measures, respectively.

In Sweden, where the stated road safety goal is zero fatalities ('Vision Zero' [201]) the widely deployed LHOVRA Traffic Signal Control System (Sect. 3.2.1.1) uses multiple loop detectors to dynamically adjust the exact signal change times in order to cater for vehicles in the dilemma zone, at the expense of traffic efficiency.

Significant work on infrastructure-based intersection collision avoidance has been carried out in the US in recent years. One system which was at the testing phase in January 2012 uses radars and lidars to measure vehicle position and speed and presents a warning on an electronic message sign [202].

3.2.2.2 Vehicle-Based Safety Approaches

Vehicle-based safety systems include in-vehicle passive and active systems, post-collision survivability systems, systems using V2V or V2I technologies and systems to detect and avoid collision with non-motorized transportation.

Passive and Active Safety Systems: Passive safety systems are deployed or are effective in response to an automobile crash; they do not act to prevent a collision. These systems protect vehicle occupants from injury once a collision occurs. Seatbelts, airbags, headrests and the passenger-safety cage may be regarded as passive safety systems. Smart airbags detect passenger weight and proximity and customize airbag deployment to the needs of passengers. Side curtain airbags for the protection of the head and side-impact airbags for the protection of the torso are among other widely-deployed passive safety systems currently available in the market.

Active safety systems help vehicles avoid collision involvement. During the 1970s greater computing power and sensor technology allowed the development of active safety systems. There is a clear evolution from the Antilock Braking System (ABS) [203], which adjusts braking force to each wheel independently to minimise wheel slip, through the Traction Control System (TCS) [204] (which can be viewed as an adjunct to ABS), which adjusts drive force to each driven wheel independently to prevent wheel spin or slip, through to current Electronic Stability Control Systems (ESC) [205] which compares the intended vehicle track obtained by measuring the steering wheel angle with the actual vehicle track which is inferred from a combination of lateral acceleration sensors, yaw sensors and wheel speeds and applies differential braking to bring the vehicle back to the intended (or a safer) track. ESC has been shown to be able to reduce the single vehicle crash rate by more than 50 % [206].

Several jurisdictions (Australia, USA, EU, Canada) have mandated that all new vehicles must be equipped with Electronic Stability Control. Vehicle manufacturers have developed additional systems based around ABS; notably Electronic Brake Force Distribution [207] and Cornering Assistance schemes [208].

All the above systems rely on sensors that monitor the vehicle to which they have been fitted. However, it should be noted that the modern high-end automobile is equipped with an array of sensors that monitor the external environment around the vehicle giving it "situational awareness", sensing potentially hazardous situations and engaging the various in-vehicle safety systems if required. Such systems include seat belt pre-tensioners, raising head rests and arming the air bag systems. Front mounted radar will inform brake assist systems if a collision is likely and lane departure systems will provide force feedback on the steering wheel if non-signalled lane departure is detected. Automated parking and parking-assist systems deserve mention but they typically have little impact on crash reduction (parking lot crashes are low speed with little risk to vehicle occupants).

Current systems integrate active-safety with passive-safety technologies for a unified approach to vehicle safety. In an example given in [209] about a situation where a vehicle approaches another vehicle at high speed from behind, the following actions may be triggered: the system informs the driver of impending danger through force feedback in the accelerator pedal and a visual warning, all windows and the sunroof are automatically closed and the braking system is pre-filled in anticipation of maximum braking, seatbelts are tightened to reduce slack and the seatbelt tensioners are activated to hold vehicle occupants in their seats, power seats are automatically adjusted to an optimal safety position, the brakes begin to automatically apply by the time the driver applies heavy pressure, the ABS system activates and the ESC operates if necessary (reproduced from [209]). These actions may be followed, if a crash occurs, with automatic notification of the crash location. Situational awareness is needed for the vehicle to detect proximity to the other vehicle in the pre-crash stage. Other safety interventions may be possible under the situation where there is V2V communication, discussed next.

V2V and V2I: Both infrastructure and vehicle-based collision avoidance systems should be viewed as necessary steps on the way to V2V and V2I based collision avoidance systems. An Austroads report [49] states that between 25 and 35 % of all serious casualty collisions (despite being Australian figures the methodology developed would be globally applicable) could be prevented if the entire vehicle fleet was equipped with DSRC-based collision warning systems. While there have been numerous trials and demonstrations [80, 81, 83, 210, 211] of DSRC based V2X based safety systems, no vehicle sold today is so equipped.

Such systems would warn drivers of an impending collision or unsafe situation (such as dangerous overtaking, blind corner, etc.). Linking to in-vehicle sensors and control systems; and potentially to road-side systems would only serve to enhance system performance due to greater situational awareness helping to decrease response time. Other safety applications include the potential to reduce conflict at unprotected railroad crossings where DSRC equipment on an approaching train would warn a car of its impending approach [212]. The message from the train could be relayed via a roadside unit in order to extend the effective warning range. Ultimately DSRC capabilities will be available in mobile devices and may provide an unprecedented level of safety to the most vulnerable road user: the pedestrian.

Several vehicle manufacturers have developed and deployed post-crash response systems (General Motors OnStar, MyFord Touch, etc.) that use the mobile phone network for communications. These systems are optional and may be available only on certain models. In Europe, however, a similar post-crash collision reporting technology termed eCall [213] has been developed and the European Commission has legislated that the phone networks in all member states must correctly handle eCall messages by 31 January 2014.

A leading factor in most crashes and near-crashes (80 % of crashes and 65 % of near-crashes) is driver inattention within the three seconds leading up to the event [214]. The study estimated that dialing a hand-held device increases a driver's risk of being involved in a crash by a factor of 3, and listening or talking on such devices increases crash risk by a factor of 1.3. The number of crashes and near-crashes attributable to dialing is nearly identical to the number associated with talking or listening, although dialing occurs less often than talking or listening. While mobile phones are the most familiar form of distraction, the National Highway Traffic Safety Administration (NHTSA) study found that reaching for a moving object while driving increases the risk of a crash or near-crash by a factor of 9, looking at an external object while driving by a factor of 3.7 and reading while driving by a factor of 3. Chapter 4 describes recent guidelines proposed by NHTSA for communications, entertainment, information gathering, and navigation devices or functions which may increase driver distraction.

3.2.2.3 Safety Systems Involving Non-Motorized Modes

In 2009, nearly half of all road deaths worldwide were vulnerable road users, including pedestrians, cyclists, and motorcyclists [215]. Here we focus on vehicle-based safety measures for pedestrians and cyclists. As in the case of systems to protect car occupants in collisions with other cars, systems to address collisions with pedestrians and cyclists can be passive or active; however, with these systems the object of protection is the pedestrian or the cyclist. Moreover, there are differences due to speeds, orientation, contact points with the vehicles, and other reasons due to which the impact resulting from motor vehicle crashes with pedestrians are different from crashes with cyclists [216].

Passive pedestrian safety measures, for example, involve vehicle-based structures such as bonnets or bumpers that expand during collision to minimize pedestrian injuries. An example is the active bonnet used by Mercedes-Benz, which, upon sensor-based impact with a pedestrian, pushes up the rear section of the bonnet by 50 mm in a fraction of a second, enlarging the impact deformation zone and potentially reducing the risk of injury. For cyclists, smart helmets that activate airbags to protect the head upon fast and atypical movements or when hit by a vehicle, screens that can be attached to the steering wheel to provide visual information about situation behind the bicycle, and various sensor-based systems that can increase the visibility of the bicycle during low-visibility conditions are examples of bicycle/cyclist-based safety systems that are either available currently or are being developed as a part of cyclist safety systems [217].

Further developments, particularly in computer vision and pattern recognition, have contributed to vehicles being equipped with pedestrian detection systems and also automatic roadside sign recognition systems [218, 219]. This sensing is a difficult problem due to the varying appearance of pedestrians (e.g., different clothes, changing size, aspect ratio, and dynamic shape) and the unstructured environment [220]. Similar issues have been noted with vision-based cyclist detection systems for driver assistance [221], with the additional caveat that the bicycle's appearance can change dramatically between viewpoints. These active driver assistance systems can detect hazardous situations involving pedestrians or cyclists ahead of time, allowing the possibility of warning the driver or of automatically controlling the vehicle by braking or swerving.

3.2.3 Dynamic Resource Management

In this section, we discuss several ways in which transportation resources and mobility services are being managed with the use of ICT. In Sect. 3.2.3.1 we examine

ICT-based dynamic vehicle fleet management strategies, followed by the increasing interconnections between transportation and energy in Sect. 3.2.3.4. This is followed by a discussion on weather-responsive transportation management (in Sect. 3.2.3.2), emergency management and crisis informatics (in Sect. 3.2.3.3) and smart cities (Sect. 3.2.3.5).

3.2.3.1 Fleet and Dynamic Resource Management

Vehicle fleet management arises in virtually all cases where there are vehicles to be operated and coordinated by an owner and/or operator. Examples are trucking fleets and commercial delivery vehicles, public transportation bus fleets, paratransit fleets, taxis, public agency vehicles including those for highway maintenance, snow removal or incident clearance, car-rental companies, car-sharing or bicycle sharing companies or vehicle fleets owned by utility companies and other private companies.

Each fleet type has its own set of operational issues and the fleet management concerns vary widely based on the nature of operations. Generally, the core problems are optimizing scheduling and routing, cost-effective vehicle relocation in case of vehicle failure, maintenance management and minimizing vehicle downtime, effective vehicle fleet utilization rates or payload utilization rates and fleet energy efficiency. Some fleets perform tasks that may be known well in advance or that are repetitive, such as vehicles making regular deliveries to food and retail stores or public transportation buses that operate on schedule along the same route. Many others operate in a demand-responsive mode where the demands for services are not known beforehand and the fleet has to be deployed or re-routed in real-time to manage them as effectively as possible. An excellent recent reference on this topic is [222].

Dynamic fleet management exploits information provided by ICT in order to improve the real-time use of transportation resources, thereby leading to real-time dispatching, routing, and re-routing of vehicles in response to changes in demand, travel time or other conditions of travel. For example, in the field of urban freight management and city logistics [223] real-time traffic conditions and time-dependent travel time estimates from sensors have been integrated into vehicle routing systems.

Scheduling taxi vehicles and drivers for large taxi operators are also complex problems for which ICT is used. A multi-agent scheduling problem is described for a very large taxi service with a fleet of more than 2,000 GPS-equipped vehicles and more than 13,000 orders per day with occasional rates exceeding 1,500 orders per hour [224]. In this system, order arrival times, pick-up locations, type of vehicle requested (van, car, special needs) and type of client (full-paying, subsidized, requiring a child-seat, pet transportation and many other variables) are unpredictable and in certain areas taxi pick-up time is guaranteed. In addition, there is an issue of driver equity in terms of allocating runs and a balance that needs to be preserved between company drivers and freelance drivers who lease vehicles from the companies. Orders and resources (clients and drivers/vehicles) are represented in the system as agents, and flexible decision-making criteria are used in a multi-criteria decision-making framework to match drivers/vehicles attributes in real-time with client needs.

In another example, two logistical measures for alleviating imbalances in the distribution of bicycles in the bike-sharing program are described in [225]. The authors applied data mining to operational data from the bike sharing schemes data to infer typical usage patterns and to forecast bike demand with the aim of supporting and improving strategic and operational planning.

Examples of Vehicle-to-Grid (V2G) management systems and paratransit (demand-responsive transit) are given in Sects. 3.2.3.4 and 3.4.3, respectively.

3.2.3.2 Weather-Responsive Mobility Services

Weather-responsive mobility services address transportation operations, management and travel assistance using real-time weather information. Inclement weather can have a range of impacts on the transportation system, including increases in frequency of crashes and secondary crashes, reduction in throughput, reduced speed, increased travel time, unreliability of travel conditions, and altered demand patterns as people either do not make a trip during bad weather or use a different route or mode of transportation. These impacts arise from physical effects that different weather conditions have on the infrastructure and environment (e.g. wetness, slick conditions, snow accumulation, reduced visibility), as well as their impacts on the driving behavior of those who may deem conditions unsafe to follow as closely as usual or to drive at higher speeds.

Several authors have examined the impacts of weather on particular roadways by investigating the changes in speed and volume of vehicle traffic under various weather conditions. Weather conditions that impact transportation conditions include precipitation, visibility, temperature, wind speed and humidity; a brief review is given in [226].

Weather information relating to transportation is available from various sources. In the US, existing road-based programs such as the Road Weather Information Systems (RWIS) and surface weather sensing systems such as the Automated Surface Observing System (ASOS) of the National Weather Service, Federal Aviation Administration and the Department of Defense are major programs. The US Connected Vehicle program supports vehicle-based weather sensing. Using a mix of vehicle-based sensors such as vehicle speed, temperature, barometric pressure, hours of operation, wiper status, rain sensor, ABS, stability and traction control, headlight status and other sources, a number of weather-related variables may be ultimately derived. These variables include ambient air temperature, barometric pressure, precipitation type, occurrence and rate, fog, pavement conditions (wet, dry, icy), boundary layer water vapour, pavement temperature, and smoke [227]. Smartphone technology is rapidly changing and, at the time of writing this book, smartphones have sensors to measure liquid contact, and after-market devices are available to measure temperature, humidity and other weather variables. In the future, these, along with other weather sensors may be in-built into smartphones, thus enabling crowd-sourced weather information where humans report weather-related information using smartphones or other devices [228].

Weather-related information continues to be one of the top pieces of information that travelers desire to have in making travel decisions. The types of information requested are whether or not to make a trip, to change departure times or modes of travel, to take a different route because of congestion, lane closures or debris accumulation, or even to evacuate from an area in the case of flooding or other weather-related hazards. Information on these factors are used in various ways for weather-responsive traveler information and traffic management. Weather-sensitive Dynamic Traffic Assignment (DTA) models which address both supply (capacity and control) and travel demand aspects of the response to adverse weather have been attempted for system-wide traffic state estimation and prediction under inclement weather [229].

3.2.3.3 Emergency Management and Crisis Informatics

ICT is used in many stages of the emergency management process, although it is more likely to be apparent in some stages than others [230]. The stages are mitigation (activities that prevent a disaster), preparedness (early warning, notification and alerts, evacuation, and recovery and rescue planning), response (actions taken during emergencies or disasters for rescue and rehabilitation) and recovery (actions that assist a community to return to a sense of normalcy after a disaster).

Rescue and rehabilitation efforts often call for field vehicle management through transportation assets that may have been destroyed, thus indicating the potential for ICT to be useful for navigation and strategic vehicle placement. For transportation organizations, route planning and traffic management on evacuation routes are major areas of concern, as is post-disaster reconnaissance of the extent and location of damage. Reverse-lane or contra-flow to enable maximum flow of vehicles, and associated variable message signs and advisories are strategies that are typically used, although such strategies are initiated only in the event of extreme emergencies, when the need arises to evacuate thousands of people in a short amount of time. Traffic control and ramp metering systems can similarly be used to create the capacity needed to evacuate large numbers of residents.

Many of these strategies focus on vehicular traffic, which does not address the question of evacuating members of carless households. As experienced in 2005 during Hurricane Katrina in the New Orleans area of the US, this can be an issue of devastating proportions, with over 1,800 deaths during the hurricane and the floods that followed. Many people who could not evacuate were poor, minority and carless. By one estimate, the number of private cars available in the New Orleans metropolitan area was enough to transport the entire population, assuming an average carrying capacity of three persons per auto, "had those with cars extended help to neighbors lacking them" [231]. While there are undoubtedly issues of prejudice, trust and other barriers to such vehicle sharing, there is also an issue of pure organizational work to manage and coordinate the evacuation needs of carless individuals, for which ICT may be helpful. But whereas social media has been increasingly used in "crisis informatics" (see below), the barriers posed by the digital divide with the evacuation of specific disadvantaged subgroups needs further study.

Social media sites currently rank as the fourth most popular source for accessing emergency information in the US [232]. Social media has been used by individuals and groups in various stages of emergencies, including to warn others of unsafe areas or situations, inform friends and family that someone is safe, support requests for information, and raise funds for disaster relief. It has been noted that whereas emergency management personnel are concerned with managing large amounts of incoming information, members of the public who are affected often experience a severe absence of timely information needed to make decisions during emergencies [233, 234]. An emerging area, crisis informatics, focuses on the interconnectedness of people and organizations through information technology—particularly social media—during crises [235, 236]). Crisis informatics focuses on redefining the top-down control and command approach to crisis information management towards community-based grass-roots strategies whereby access to information and knowledge is created and disseminated by citizens using a mix of methods to communicate and disseminate information [235].

Crisis informatics also focuses on new roles and functions emerging as people, including both those who are within the affected region and those outside, go online or to mobile ICT to provide, seek and broker information. For example, people self-organize in response to disasters to conduct search and rescue, administer first aid, provide shelter, and participate in other activities by which rehabilitation and recovery may be facilitated. Problems considered within the area of crisis informatics include understanding the means both for disseminating information and for connecting with other local people during times of crisis [233], and developing a data infrastructure that manages information from microblogging during disasters [237].

3.2.3.4 Energy, Smart Grid, V2G and Electric Vehicle Information Systems

Connecting transportation to energy management has received significant attention recently as countries around the world seek to reduce the use of fossil fuels and minimize carbon impact. The focus of research has been on two broad areas: Alternative Fuels/Alternative Fuel Vehicles and energy management and use practices. In this section, we focus on the latter and review the major developments regarding ways in which information technology has been used to leverage energy management and low-carbon practices in the transportation sector.

One area is intelligent management of Electric Vehicle (EV) charging. For example, on the island of Bornholm, Denmark, EVs are integrated into the local power grid, which relies heavily on renewable wind power. The Bornholm electricity grid uses a network of public and personal charging stations and integrated technologies to manage the charging of EVs as well as load balancing, billing, and other functionalities [238, 239]. To the extent possible and allowed by consumer preferences, EVs will be charged when wind is generating excess power, and, conversely, EV charging will be slowed or delayed when the wind stops and energy production is diminished. Bornholm has been noted as a useful test case for assessing ways in which renewable energy can be combined with EVs since, as an island, its electricity grid is self-contained and isolated, making it easier to manage outcomes and analyze impact [238].

Vehicle-to-Grid (V2G) systems utilize the stored energy in the batteries of parked EVs to deliver electricity back to the electrical network when requested by the network operators. By one estimate, the power capacity of the internal combustion passenger vehicle fleet in the mid-1990s in the US had over 10 times the mechanical power of all current US electrical generating plants and is idle over 95 % of the day [240]. Alternative fuel vehicles can be used in different ways to assist in power storage and generation (for example, battery vehicles could be used for storage while fuel cell and hybrid vehicles could be used for generation). Most recent literature on V2G has focused on EVs. It is noted in [241] that V2G appears to be unsuitable for baseload power where electricity supply is provided constantly round-the clock since baseload power can be provided more cheaply by large generators, as is currently the practice. Rather, the greatest near-term promise for V2G is for quick-response, high-value electric services to balance constant fluctuations in load and to adapt to unexpected equipment failures.

One motivating factor in examining the connectivity of the power and transportation sectors are Electrical Markets (EM), which are a consequence of the deregulation of electricity production and use, where power suppliers and consumers are free to negotiate the terms of their contracts. For example, policies relating to the US Smart Grid, (the stated goal of which is to deliver electricity from suppliers to consumers using digital technology to save energy, reduce cost, and increase reliability and transparency) foresees a future where there are producers/sellers of power and electric companies, which are buyers, thereby encouraging market designs that compensate consumers for producing energy. Such integration is being considered within a larger framework in which individual households, equipped with solar panels and household wind turbines, as individual household producers of energy within a green energy infrastructure, can sell excess renewable energy generated to the city to generate income. In [241] Kempton and Tomic projected the value of a V2G (grid-enabled) vehicle that provides ancillary services to the grid, at upwards of $3,000 annually.

ICTs may also facilitate the idea of personal carbon trading, where a total cap would be set on all carbon emissions generated from fossil fuel energy used by individuals within the home and for personal transportation. Each person would receive a personal carbon allowance allocated on an equal per capita basis. Individuals would use their allowance when purchasing fossil fuel; if allowable carbon limits are exceeded, then the individual would need to purchase additional carbon units; if there is a surplus, carbon units could be sold. For the transportation aspect of a person's total carbon emission, there would be a need to track the level of use by mode, time-of-day and so on. Such a system has been proposed for London [242]. In the US, attention has been paid in recent years to the idea of a driving or mileage-based tax where drivers would be taxed based on how much they drive in contrast to fuel tax.

A related area of research is on carbon emissions resulting from the energy use of vehicles. Research in vehicle communications opens up the possibility of dynamic carbon emissions trading. An example with pollution trading is given in [243] where a distributed m-commerce network Vehicular Cap-And-Trade (VCAT) is proposed for vehicle owners to buy and sell pollution rights. Each vehicle would receive pollution rights and be penalized for exceeding them. Drivers could trade any unused pollution rights with businesses as well as other vehicles. The network would monitor real-time pollution and traffic levels, manage pollution accounts, and facilitate transactions between vehicles. Future scenarios may involve dynamic trading of carbon allowances in such a connected vehicle environment.

Some recent examples of connecting transportation to energy overlaps with vehicle fleet management strategies (see Sect. 3.2.3.1). One example is the safe and contactless charging systems for on-street charging stations being developed by MIT for the Mobility on Demand (MoD) ultra subcompact vehicles concept [244]. Other examples are deployments of emerging smart city initiatives (see Sect. 3.2.3.5), for example, the Bornholm, Denmark example given above and South Korea's U-Eco City Initiative which explicitly integrates city management/operations and citizen services with green technologies [245].

Personal energy demand management is another area of interest, particularly in measuring and providing feedback on carbon footprint and energy use. Many vehicles now have devices that display carbon output, and there are now websites that provide not just trip plans but also the carbon impact of the trip. CarbonRecord is a mobile-social application that is designed to enable individuals to track their daily vehicular carbon emission through an Android app and share them on a social network, Facebook, with the goal of raising awareness about vehicular carbon emission [246]. CO2GO is another (iPhone-based) app [247] that uses accelerometers, GPS and other embedded sensors and deploys a context-aware algorithm to calculate in real-time the carbon emissions of the user, by automatically detecting their transportation modes and tracking the distance covered.

Eco-driving, a set of driving behaviors that reduce fuel use and carbon emissions, is another approach. Eco-driving includes accelerating smoothly, coasting to a stop and not idling too long while driving, and car maintenance practices such as inflating tires properly, using the appropriate oil and changing air filters regularly. Dynamic eco-driving technology provides real-time driving advice such as recommended driving speeds, optimal acceleration and deceleration profiles, and alerts, so that drivers can adjust driving behavior or take certain driving actions in order to save fuel and reduce emissions. The most sophisticated versions give visual or oral instructions on how much pressure to apply to the accelerator pedal. Social sharing aspects result in uploading records of driving behavior from vehicles to Web sites where participants can compete with others to be the most efficient driver. For example, Youldeco is a social network which is based on rewarding eco-friendly driving by making it competitive and game-like [248].

Various other aspects of the driving situation have been capitalized towards the goal of eco-friendly driving. For example, [249] describes an intelligent speed adaptation system. Another application—Eco-Signal Operation reported in [250]—takes advantage of traffic control information about signal phase and timing available through DSRC to possibly provide alerts to individual vehicles entering the DSRC range that have little or no chance of traversing the intersection before the signal turns red. Once such information is disseminated, vehicles can start to coast down to a full stop instead of continuing at their current speed and braking hard at the stopline.

While ICT-based navigation and travel assistance, as described in Sect. 3.4, has the potential to improve energy efficiency and environmental sustainability by avoiding congested routes, and supporting public transportation and non-motorized travel, it needs to be kept in mind that to date, ICT impacts on such travel behavior has been limited. These issues are discussed in greater detail in Sect. 4.4.

3.2.3.5 Smart City, u-City, e-City, Digital City and Transportation

The previous sections described several ways in which other sectors (energy, weather, emergency management) can be combined with transportation. Also as noted in Chap. 1, there are many terms including smart city, e-city, u-city, digital city and perhaps others, where transportation is part of a mix of overall strategies to incorporate ICT-based intelligence into the fabric of cities. These concepts are a continuum that emphasize different functionalities and integration aspects. They leverage investments in broadband, wireless technologies, smart grids and connected devices, and the increasing ubiquity of sensors in urban infrastructure.

Several opportunities and challenges are presented by the ideas of integrating different aspects of government and society, including cost-effectiveness, economic development and overall improved quality of life. The integration of the technical subsystems within cities has stimulated innovations in smart urban infrastructure, M2M communications, Big Data analytics, dynamic resource management, knowledge discovery for improved operations, open data policies and indicator development metrics for measuring the level of "smartness" of cities. Several examples of smart cities currently exist, albeit at various stages of development and scale. A common early example in such a "system of systems" approach is the integration of transportation and energy systems through integrated green technology solutions, as described in Sect. 3.2.3.4.

At the same time, there are challenges. For example, it has been noted that the process by means of which a city can transform itself into being "smart" can be involved and can take a long time [239]. The authors also note that the process may be motivated by reasons as diverse as recovery from a natural disaster, large-scale special events, or a visionary city leader who understands the role that technology can play in civic engagement and economic development, and channels their energies into its development. The concept has also been considered to be fuzzy with multiple meanings. Some authors have focused on cities that use ICT to "speed up bureaucratic processes and help to identify new, innovative solutions to city management complexity in order to improve sustainability and livability" [251]. Others have noted that a city is smart when "investments in human and social capital and traditional (transportation) and modern (ICT) communication infrastructure fuel sustainable economic growth and a high quality of life" [252].

Overall, current city initiatives have been noted to be too technology-focused at the expense of ignoring the human capital aspects of cities and for using narrow or irrelevant metrics for evaluation. Nevertheless, transportation and mobility services will continue to be interconnected with other urban systems and services although at different rates and scales, and potentially to the increasingly ubiquitous information society.

3.3 User-Generated Content: Involving People in Mobility Management

In Chap. 2, we noted that UGC can occur in two modes: proactive UGC, by supporting or enlisting users to generate primary data on events, concepts, ideas or activities of interest; or retroactive UGC, by processing secondary user-submitted data that is published using social media tools such as Web 2.0 tools, blogs, microblogs and so on. In this section, we will discuss different ways by which proactive content and retroactive content can be generated and types of mobility services that can be built using UGC.

3.3.1 Proactive User-Generated Content

Overall, citizen-based approaches can be categorized as contributory, collaborative, or co-created [253]. The use of sensors and social media tools by ordinary individuals to generate transportation-related information has been motivated by several synergistic threads with different origins. The first is the area of citizen participation in transportation planning. The second is the area of citizen science and journalism, where volunteer interest-groups armed with sensors and measurement devices collect information of interest to them; a subset of this stream morphed into what came to be known as participatory urbanism. The third emanated from ITS, where, for example, cars equipped with in-vehicle navigation devices, at the discretion and participation of the drivers, would provide speed information as vehicle probes and receive navigation directions as consumers of information.

Citizens using ICT for mobility sensing are supported by various enablers who provide the technology or decision-support basis to support their activities. Such enablers include: (1) *Open-source software and platforms* some of which are playing an increasingly important role in mobility informatics; an example is Hadoop, which allows computationally intensive analytics and mining of massive amounts of structured and unstructured Big Data for UGC to be carried out in "the cloud"; (2) *Mashups* which integrate disparate data sources and deliver aggregated information to the consumer, thereby allowing transportation users to come closer to complex arrays of data. (3) *Open Data portals*, particularly government entrepreneurs who are facilitating open access to data that were previously inaccessible to ordinary citizens. (4) *Community decision support* or tools which support community decisions, particularly Public Participatory GIS (PPGIS) developers and social media for community visioning and participation. (5) *Digital civic entrepreneurship or civic hacking* is another thread in the institutional mix helping organizations and actors that push data out to citizens with a myriad of goals to support increased awareness and information on communities, civic engagement, community participation and user submissions.

There are several criteria by which proactive information generation can be categorized: by type of content, type of compensation, or type of motivation behind information generation.

Type of Content: There are three types of content that may be proactively generated by users: (1) idea generation, feedback and problem solving content; (2) human computation for tasks that are difficult for machines to perform; and (3) sensing information content.

1. **Idea Generation, Feedback and Problem Solving**: Strategies by means of which ideas may be generated and envisioned have long been a part of transportation related decision-making. Such strategies include brainstorming [254], Strength Weakness Opportunities and Threats (SWOT) analysis (attributed to Albert Humphrey of Stanford University) and others such as Future Workshop (a technique for envisioning futures and which includes a fantasy phase where participants try to work out a utopia, to draw an exaggerated picture of future possibilities) [255] and Scenario Approach involving scenario analysis which uses certain techniques and steps leading to construction of quantitative scenario pictures of the future [256]. A range of methods that fall into the areas of soft operations research have been used in transportation planning to solve complex problems with the assistance of citizens. Two well-known methods are SODA [257] and Soft Systems Methodology [258].

   The use of information technology to extend these ideas has led to a plethora of approaches for citizen problem solving, plan and design sourcing, voting on projects, and sharing of ideas for projects. In 1996 Graham [259] argued that the Web will "generate a new public sphere supporting interaction, debate, new forms of democracy and 'cyber cultures' which feedback to support a renaissance in the social and cultural life of cities". Developments in this area have been rapid and there are currently several ways in which social media augments the idea-generation and feedback processes involved in traditional transportation planning. The power of social media, of course, is that a much larger (Internet-scale) audience can be drawn into problem-solving and idea-generation related to transportation and mobility.

   There are, however, important differences and potential constraints on such power. First, many transportation problems tend to be very localized in nature (for example, only people residing in a certain area possibly have enough local knowledge to come up with ideas that are congruent with the needs of a local transportation project), thus the input from an Internet-scale audience may not be directly relevant. There may, however, be examples of solutions generated for similar problems in other areas not documented in the formal literature or not available through professional planning networks that may be offered by people from those other areas. Second, although the use of information technology opens up the process to a larger group of people who may not be able to commit the time and resources to be involved in traditional processes, issues of information technology access, digital divide and outreach may still preclude fuller levels of participation. Other limitations are discussed in Sect. 4.2.4.2.

2. **Human Computation and "Volunteered" Work**: A second way in which users may generate content is by performing tasks which are natural for humans but difficult for machines to automatically carry-out. Human computation is a means of solving a computational problem by dividing the computational task between humans and computers. Often used examples are CAPTCHA [260], which is a computer generated challenge-response test to distinguish humans from computers using a common sense problem and reCAPTCHA [261], which is a novel CAPTCHA that attempts to improve the Optical Character Recognition (OCR) quality in digitizing books for the Internet Archive.

Although the approach has not been applied in transportation applications, there are several ways in which humans can be used to perform tasks that assist in transportation decision-making, particularly where objective data-based metrics are not accurate enough to comprehensively judge the quality and level of service of transportation programs, and where perceptions and judgments of large numbers of people would be useful. An example where such an approach could be used is in Spatial Decision Support Systems, where large numbers of users could be given the task of visually recognizing and judging the quality of transit stations from online photographs of transit stations that have been user-submitted and are geotagged. The goal would be to build realistic databases relating to the quality of the built environment surrounding transit stations. Here, users could be given the task of recognizing the quality of a random station from its photo when they upload and tag their own photo of a transit station. Another example is to use humans to rate the quality of service in transit routes to identify routes that consistently provide poor service.

3. **Sensing**: A third approach by which users may generate information is data collection by sensing. Proactive sensing modes can be disaggregated into three different types, depending on the level of decision-making needed on the part of the user:

   a. *Participatory sensing*: In participatory sensing, users opt into a sensing system and *actively* report events witnessed or share mobility experiences. The sensing system may be a web-based system through which users subscribe and submit information, a mobile device system, or a hybrid of the two. Participatory sensing may involve several types of activities. Examples include geographic information submission regarding interesting places, routes and transportation options; community monitoring to report potholes, dangerous intersections and the like; reporting of road incidents, transit delays, special events, and a myriad of other ways in which individuals subscribed to a sensor system submit information. Several terms have been used to refer to data created through participation or volunteerism; for example, volunteered geographic information is the creation, assembly, and dissemination of geographic data provided voluntarily by individuals [262]. Several examples currently exist including WikiMapia, OpenStreetMap, Google Earth and Google Map Maker.

b. *Opportunistic Sensing*: In opportunistic sensing, users enable their wearable or in-vehicle location-aware devices to automatically track and *passively* transmit their movements. A number of real-time automotive tracking applications currently measure speeds yielding data on congestion, incidents and the like. Waze is an example of a peer-to-peer application that collects travel-time information from users' smartphone-based GPS. The point data and user reports are uploaded to a central application, as well as exchanged to other local users in a peer-to-peer network. Another example is Biketastic, which uses GPS-based sensing on a mobile phone application to enable sharing of cyclist experiences with one another. Cycling speed can be inferred from the GPS sensing capabilities, while the microphone and the accelerometer embedded on the phone are sampled to infer route noise level and roughness.

c. *Ad-hoc sensing*: Early deployments of transportation information systems are likely to depend greatly on the activities of users who are not in any way a part of organized sensing systems but who report on travel and transportation conditions to larger systems operated by government agencies or private companies. Examples are cell phone users who call in to report traffic incidents or road-related weather conditions. Customer feedback of transit systems, such as delay-reporting through microblogs, is another example.

Compensation: Aside from the type of content, another criteria to distinguish different forms of information is whether there is a payment involved for the task; for example, the crowd may be paid (rented or hired) or unpaid volunteers. Payments may be by means of incentives or actual money.

Additionally, payments for one-time tasks such as idea generation or problem solving may occur through contests and competitions where challenges are issued to generate transportation ideas and meet infrastructure design, software, service solutions and other needs. Payments may also be on a continual basis.

Type of Motivation: Authors have distinguished different motivating factors behind proactive content generation. The purpose for which UGC is generated may be social-driven (as in sharing via location-based social networks or volunteering information by user-based communities), purpose-driven (as in commercial LBS supporting car or multimodal navigation) or a hybrid of the two. These ideas are given in [263]. Pay, altruism, enjoyment, reputation and implicit work may be motivating factors behind content generation and human computation [264].

3.3.2 Retroactive User-Generated Content

In retroactive UGC, content posted online using social media such as social networks, microblogs and other forms of UGC are processed by analysts to create an information source for transportation needs and experiences. Web data mining is an area which deals with discovering useful information from the web. Web mining tasks can be categorized into three types: web structure mining, web usage mining and web content mining [265]. Web structure mining discovers knowledge from hyperlinks, such as important web pages, which is a key technology used in search engines or communities of users who share common interests. Web usage mining (WUM) discovers user access patterns from web usage logs.

Web content mining (popularly known as web scraping) extracts useful information or knowledge from Web page content. Structured data (numbers in a database, for example), semi-structured data (where the structure is irregular, heterogeneous, implicit or of non-rigid structure) or unstructured text data which typically results in UGC, and from which information can be extracted using keywords, word frequencies and Natural Language Processing can be the object of the web content mining activity. In [266] Sasaki, et.al. gives an example of extracting train status information from tweets. The solution presented consists of a keyword based analysis of tweets referring to the status (keywords include "stopped", "delay" or "suspend") of a specific train line searched and collected via the Twitter Search API [267]. A set of heuristic rules help to further refine the identification of train status tweets. To ascertain information reliability, train status information is determined to exist if the number of positively identified tweets posted within the specific time window exceeds a certain threshold.

Geotagged data resulting from the process of assigning geospatial context information ranging from specific point locations to arbitrarily shaped regions is of prime significance for retroactively extracting transportation and mobility intelligence from UGC. Pfoser [268] noted that people respond to data in the way it is presented, geospatial coordinates and routes for example. He then points out that this is not the way geographical information is passed from person-to-person. The phrase "A is close to B" is much more likely to be heard than a string of grid references. The challenge is to interpret this very human data and link it with more precise sensor-derived location data. The former, textual interpretation, sits firmly in the field of Natural Language Processing (NLP) and has been progressed by Hornsby and Li [269], for example; and the latter in information fusion (of which the perhaps more familiar sensor or data fusion are subsets) which is a principled way of combining data from multiple sensors to yield better results than any single sensor could produce on its own [109]. Temporal reasoning in natural language is equally challenging [270].

A related technology is content aggregation, which automatically grabs information of interest from multiple online sources, filters them by using persons (editors and moderators) who determine the value of the link; or based on user ratings about whether they "liked" the content, thereby connecting users to sites that discuss information of interest. These approaches are likely to lead to different ways of identifying transportation concerns of importance to decision-makers.

Rich but less-examined sources of content are question-and-answer databases relating to mobility and opinion-sharing available in community wikis and neighborhood management systems. Although these systems are not sensing systems in the traditional sense, users submit information voluntarily and are likely to serve as the most comprehensive and distributed "sensors" for determining the state of the transportation system.

3.3.3 Issues to Consider in Designing Sensing Programs

Specific applications may be more strongly supported by one type of sensing system compared to another. A proactive sensing program needs to have a critical mass of users for community appeal and engagement since community engagement and interaction is a part of the appeal of belonging to human sensor systems. Where information is to be monitored on a continual basis, as in the case of highway network speed generation, opportunistic sensing makes more sense. Proactive sensing may be more useful in generating information on intermittent events or submitting user experiences which may be difficult to detect using machines. But it has been noted that proactive sensing places demands on users who are involved (e.g., prompting via their device GUI, keying or voicing information, enduring interruptions while driving or walking) which may restrict the pool of willing participants [271].

Models of proactive UGC raise other issues some of which are legal, such as liability in the event of injuries. There is also the question of how the participatory sensing community forms, grows and sustains itself over time. This question has been examined in the case of development of online communities in terms of interactions in a multi-agent setting and the emergent intelligence that occurs at the collective level [272], members' attachment to online communities due to group identity formation and interpersonal bonds [273], reputational influences [274] and the relative anonymity, selective self-disclosure, ability for self-expression, creation of "virtual identities" and various other traits [275–277]. Strategies such as reputation and trust management [278, 279] and incentives [280] could be useful in improving UGC data quality and overall participation.

In order to increase the scale and diversity of applications that may not otherwise be supported, it may appear that a strong motivation exists for opportunistic sensing in transportation because they automatically shift the burden of supporting an application from the custodian to the sensing system. However, this issue may be very subject-dependent since detecting patterns from opportunistic sensing may be difficult. For example, road incidents may be more easily detected from driver mobile phone calls or from drivers in stalled traffic entering the information into a mobile sensing system than from an analysis of highway sensor data. However, this also places the burden on the drivers to accurately detect and report the incident. In both cases, the ability to discriminate a slow-down in traffic due to ordinary "recurring" congestion versus "non-recurring" congestion may be challenging.

There is also a need to establish systems for the evaluation of the quality of UGC data in terms of its validity and accuracy. While the quality of specific types of transportation UGC (such as probe vehicle data quality) has been researched for a while [281, 282], such content has become much more diverse in recent years as a result of proliferation of mobile phones, Web 2.0 and social media. Ways in which traditional methods of data quality assessments may need to be augmented for UGC quality evaluations are discussed in [283, 284]. As noted in Chap. 2, sensor fusion is a major challenge with new problems highlighted by each new type of sensor that yields data for mobility intelligence. Integrating heterogeneous UGC with traditional transportation sensor databases poses special challenges to sensor fusion technology [285, 286]. Moreover, selection biases and lack of data representativeness in UGC have been only very recently considered [287, 288] and needs to be explored further particularly in terms of geographical and demographic representativeness.

Aside from the technical challenges of mining user-submitted data to extract mobility intelligence, various legal and ethical issues may arise with retroactive UGC. These include privacy, copyright infringement and other issues that are discussed in Chap. 4.

3.4 Technologies for Personal Mobility and Accessibility

We discuss three groups of technologies: those for travel information and location-based services (in Sect. 3.4.1), mobile health and wellness technologies (Sect. 3.4.2), and finally, assistive and related technologies which support mobility needs of those with physical and functional disabilities (in Sect. 3.4.3).

3.4.1 Travel Information and Location Services

Technologies for personal mobility and accessibility support travel-related decisions. Travelers have a plethora of information needs regarding when, where, how and with whom to travel. Travel information needs vary depending on whether the travel environment is familiar or unfamiliar to the person, whether there is an expectation of unreliability in travel conditions (e.g., due to congestion or bad weather), whether emergencies are involved, and many other factors. Travel itself may be stimulated by information regarding destination opportunities (social, economic or other).

Technologies for personal mobility are usually called Advanced Traveler Information Systems (ATIS). They are also a leading aspect of LBS. At the current time, personal mobility services range from those that deal strictly with travel and transportation (such as geospatial resource discovery or navigation and route-finding from point A to point B) to those which support multiple functionalities (where mobility is integrated with social, environmental and personal goals and preferences). The types of information requested include location-based "where-am-I" and "what's-near-me" types of information, travel directions and routing information, congestion-related information, next bus arrival, and so on. For example, a scan of third-party applications available over the Internet and in mobile marketplaces show a wide range of specialized apps which support specific personal travel and mobility activities. A far from complete list includes those generating carpool opportunities, nearest parking and cheap gas stations, transit service alerts or road congestion and incident alerts, marking and navigating back to parking spots and coordinating and sharing parking with neighbors.

Other examples are on tracking cyclists and pedestrians, information on personal health and calories burnt, and eco-feedback relating to pollution emission and carbon footprint. Data on unique travel and mobility-related information needs and human/collaborative responses to such queries can be gleaned from information on Internet sites, such as question and answer databases, comments posted to blogs or newspaper reports and other reports on the web.

Finally, methods of information delivery can markedly differ. Information may be available from infrastructure-based solutions such as Variable Message Signs along highways or in transit stations or depending on the context, may be available through in-car navigation devices, desktop computers or via handheld mobile devices using specialized apps or social media.

3.4.1.1 Types of Personal Mobility and Accessibility Services

In terms of the content of mobility services, there are five major categories: (1) finding places, persons or objects of interest (resource discovery systems); (2) how (and/or when) to get to locations (navigation systems); (3) how to fit travel into daily activities and schedules (scheduling systems); (4) how to travel in ways which satisfy social goals or to generate new social networks (Location-Based Social Networks); and (5) how to travel in ways to meet personal (eco-friendliness, health and wellbeing, financial, etc) goals (Location-Based Personal Productivity Systems). It is possible that many of these features are integrated into one system.

Resource Discovery Systems: A common mobility information request is geospatial resource discovery (eg., "where is the nearest gas station?", "where is the nearest Chinese restaurant that offers buffet lunches for less than $10?", or "where is the nearest bus stop going downtown so that I can walk for 10 min?"). For a mobile user, such requests require a mobile device, positioning, map-matching, a content database, communications network, and spatial buffering and routing capabilities. Such queries have stimulated research in several aspects of Geographic Information Retrieval (GIR), which have Information Retrieval technologies at their core but which in addition emphasize spatial and geographic information.

While many of these capabilities have been developed for spatial data management and for the web environment, searches through mobile devices and positioning systems have triggered a need for efficiently finding information about objects in the vicinity of the device. Advancements in GIR supporting such personal mobility queries include geographic data management with improved architectures, and data structures and related techniques particularly in the areas of metadata or structured sets of data that describe other data. The objective of these strategies is to improve knowledge about described geographical resources, to answer questions and locate and use data resources, to aid web searches, and also to facilitate interoperability in distributed environments. For a recent series of papers on the topic of geospatial resource discovery, see [289]. Additional lines of work which support geospatial resource discovery for the traveler include geospatial search [290, 291], geographic information extraction [292], geospatial web crawling [293], ontology development [294] and spatial knowledge acquisition and cognition [295]. In the sphere of transportation, management and querying in the context of mobile objects and mobile resource discovery are also areas supporting personal mobility technologies.

Navigation and Routing Systems: A second functionality is navigation and wayfinding (for example, "what is the best way to get from here to Point B by car?"). One aspect of this problem is the calculation of the shortest path between two points (a survey article is [296], while a recent reference targeted to problems in car navigation systems is [297]). Significant advancements have been made in the area of route planning during congestion. This problem involves predicting travel times during congestion, so that accurate trip time from user-requested origin to destination can be given.

One line of work has been on traffic prediction (predicting travel times or traffic speeds) using in-road and in-vehicle sensors. Statistical forecasting and data mining methods are used for this purpose, using time-series modeling, Kalman filtering, neural networks, non-parametric regressions, genetic algorithms, fuzzy logic and several other approaches [298]. The basic problem is this: given measured speed at a location, what is the forecasted speed at a future time (say, 5, 10 or 15 min into the future)? The shortest travel time route can be calculated once network link costs have been updated to reflect the travel time (or speed) at the (clock) time which the vehicle is expected to enter the link. Another approach to predicting the short-term status of transportation networks is Dynamic Traffic Assignment (DTA), which are mathematical optimization or simulation models which simulate the dynamics of traffic networks [299, 300]. DTA methods are used for a number of real-time and planning purposes. For real-time purposes, DTA aims at determining the time-dependent traffic network link flows given a set of travel origin-destination (O-D) zone-based traffic demands for a given time period, by following certain behavioral assumptions about travelers and physical models that govern the flow of traffic on links and through intersections.

By some accounts, traffic incidents such as traffic crashes and vehicle stalling lead to unexpected changes in travel times and speeds, implying that a user should be routed around such trouble-spots. Research into traffic incident effects has focused on AID algorithms where the occurrence of a traffic incident is detected using sensors (see Sect. 3.2.1.3). Other research has focused on estimating trip times during special events, inclement weather and construction.

Several recent research findings regarding mobile searches have implications on personal mobility and assistance technology development. First, large mobile search queries request several bundled pieces of information that include basic routing and navigation within searches for location-based and activity information. Second, although the number of mobile search queries has exploded, the fraction of queries relating to travel, mobility and transportation within these mobile searches remain small. Third, many searches for location and navigation information are often done using mobile devices in static, familiar settings. The implication is that basic navigation-type information should be blended together with other functionalities that make mobility information socially relevant to the user.

Personalized Itinerary Scheduling Systems: An area of personal mobility support is trip, ride, activity or itinerary scheduling (for example, "I would like to coordinate a ride with someone" or "I would like to find someone to walk with from the bus stop to home"). Whereas one line of work in this area has focused on ways to seamlessly connect travelers with shared transportation, another stemming from travel behavior research has focused on ways in which travelers schedule their activities and travel. A method has been proposed on iteratively designing technology to support and grow mobile social ridesharing networks in particular areas; this tool improves convenience and usability of ridesharing by allowing people to easily contact persons in their extended ride-share social network through mobile phones for a ride [301].

Some of this work rests on inferring similarities in mobility traces and matching travelers with similar itineraries. One example is the extraction of mobility profiles of individuals from raw GPS traces and to study criteria to match individuals based on profiles for carpooling [302]. Linking such capabilities with dynamic models of activity planning and scheduling [303] would likely result in personalized itinerary scheduling systems that link travel plans to sustainable modes of transportation.

Location-Based Social Networks: Given the proliferation of social networks, it is not surprising that Location-Based Social Networks (LBSN) have emerged as an important aspect of personal mobility and accessibility services. LBSN activities include geotagging (adding geographical identification metadata to various media such as photographs, video, websites, microblogs or curated content such as RSS feeds) and location-sharing. Key to the development of LBSN is context-awareness: "where you are?", "who you are with?" and "what resources are nearby?", i.e., exploiting the current location of the user, their social network and the availability of resources of interest in the user's surrounding. An LBSN not only means adding a location to an existing social network so that people in the social structure can share location-embedded information, but also consists of the new social structure made up of individuals connected by the interdependency derived from their locations in the physical world as well as their location-tagged media content, such as photos, video, and texts [304].

Connecting traveler information with recommender systems has augmented the ability of users to find location-based information of interest. Recommender systems emerged as a concept in the 1990s to harness the opinions of millions of people online to help find useful and interesting content [305]. These systems are information filtering and decision support tools, which automatically and proactively select personalized resources that best suit the needs and preferences of each user by using a variety of techniques.

One commonly used approach for making predictions of what to recommend to a user is collaborative filtering. In collaborative filtering, users are recommended items that others with whom they are similar (in terms of explicit user ratings of items made in the past, or implicit ratings derived from browsing history, purchases and so on) have selected. In such an approach, the recommendation (prediction) is made based on the similarity of the user profiles and does not explicitly need to consider the attributes of the items themselves. The method is called collaborative filtering because personalized recommendations can be made for selected items out of a large range of items and the users implicitly collaborate with each other. Other techniques are content-based recommendations which suggest items which are similar to those a given user has liked in the past or may be knowledge-based recommendations or a hybrid of the two former approaches.

Recommender systems have been studied in various contexts relating to the mobile environment. For example, in the area of mobile commerce (mCommerce), a peer-to-peer mCommerce collaborative filtering-based recommendation system is given in [306]. Route recommendation systems have been studied in several contexts. The case of recommendations of taxi routes to taxi drivers for successful taxi-pickups that are predicted based on their past performances, in terms of revenue per energy use, in a way that the potential travel distance before reaching a customer is minimized is examined in [307], while [308] considers the case of event location with personalized (real-time) public transportation and pedestrian route recommendations.

Developments within LBSNs enable understanding locations on the basis of collective social knowledge of users and travel recommendations that can be given using such understanding. One personalized recommender system learns the user's interests from the user's personal location data based on GPS trajectories, and suggests locations to the user that match their preferences [309]. The system uses the times that user has visited a location as the user's implicit ratings for that location, and predict the user's interests regarding places they have not yet visited by considering the user's location history and those of other users. Locations with high ratings that match the user's preferences can be recommended.

Location-Based Personal Productivity Systems: Finally, personal mobility and accessibility services have expanded to give people the ability to fulfill personal goals such as environmental awareness, personal health-related goals and so on while traveling. These may occur, for example, through eco-feedback or through feedback and awareness regarding health and wellness. Social systems using persuasion to integrate mobility with health, sustainability, gaming and personal informatics are increasingly finding their way into mainstream use. Examples relating to eco-feedback and eco-travel applications were given in Sect. 3.2.3.4 while health-related and persuasive technologies are given in Sect. 3.4.2.

3.4.2 Mobile Health and Wellness Technologies

ICT has been playing an increasingly important role in connecting transportation to health. Here we review recent developments in mobile health and wellness informatics, and ubiquitous health monitoring through environmental health systems and in-vehicle systems. These applications have been enabled by wearable sensors, mobile phones equipped with sensors, web cameras, biometric sensors and online social networks. In-vehicle environments represent another area in ubiquitous wellness informatics.

Mobile Health and Wellness Informatics: As noted in Chap. 1, active transportation policies facilitate physical activity by increasing opportunities for walking and cycling. Transportation policy activities have focused on infrastructure-based strategies to improve the quality of the mobility experience, such as building sidewalks, bicycle lanes and tracks, as well as on traffic control that is responsive to non-motorized transportation. Education and enforcement activities to support safe travel by these modes are also a part of active transportation policies.

Modes of Physical Activity Support: ICT support for physical activity occurs in two modes: opportunistic and structured physical activity. The extent to which applications can be used to build walking or running into people's daily schedules and travel itineraries is an important area of research. Such applications target opportunistic physical activities, where a person incorporates activities into her normal, everyday life to increase her overall level of physical activity [310], in contrast to structured physical activity, where a person elevates her heart rate for the purpose of exercising.

Examples of opportunistic physical activities include walking instead of driving to work, taking the stairs instead of the elevator, walking to the next bus stop or parking further away from the office. Such travel activities, derived from daily travel behavior, are in fact the target of active transportation policies. ICT efforts on opportunistic physical activities have focused on three major elements: use of mobile devices to provide information at the right place and time so as to be able to substitute motorized mobility with physical mobility; use of social networks to build support and share experiences with friends and family; and use of persuasive technology principles to motivate an active lifestyle. Much of ICT-based support for physical activity has leveraged sensors in smartphones such as the gyroscope, accelerometer and compass.

User-Centered Design and Social Design Principles: By incorporating such principles, [310] proposed four design principles for a prototype mobile phone-based fitness journal to encourage physical activity, particularly step count: giving users proper credit for activities, providing personal awareness of activity level, enabling supportive social influence, and considering the practical constraints of users' lifestyles. The prototype consists of a pedometer, a mobile phone and software that runs on the phone. The system allows interaction with other fitness buddies in the social network by sharing step count information.

In the BikeNet mobile sensing system [311, 312], bicycles are outfitted with specialized remote sensors (custom Tmote Invent motes), and sensor-enabled mobile phones are attached to cyclists' helmets. The system requires the integration of heterogeneous sensing platforms into a single architecture with a common set of interfaces. The data includes the cyclist's personal data such as heart rate and galvanic skin response, the cyclist's performance such as wheel speed, pedal cadence and frame tilt, and location-specific data such as the sound level, carbon dioxide level and proximity of cars. The sensor data are uploaded when the bicycle comes within radio range of a sensor access point and can be shared with other cyclists in real time. Information may also be shared bicycle-to-bicycle via short-range radio or indirectly through neutral storage and aggregation devices.

Persuasive Technology Principles: Persuasion to motivate healthy behaviors is a step further from monitoring. The study of computers as persuasive technologies is termed "captology", based on the phrase "computers as persuasive technologies"; such technologies are systems, devices, or applications intentionally designed to change a person's attitude or behavior in predetermined ways without using coercion and deception [313].

As a framework for thinking about the roles that computing products play, a functional triad was proposed [314]: as tools, as media, and as social actors. As a tool, interactive computing technologies can be persuasive by making target behavior easier or more efficient to do, leading people through a process, intervening at the right time and place with a suggestion, or performing calculations/measurements that motivate and engage. An example of a health-related mobility tool is to keep track of how many steps need to be walked in order to complete a daily travel itinerary that meets a daily physical activity goal. Persuasive technology principles are being leveraged to motivate rather than simply monitor healthy behavior such as physical activity using mobile devices to present just-in-time information at critical decision points. As media, technologies can be persuasive by allowing people to explore cause-and-effect relationships, providing people with sensory experiences that motivate and convey messages in vivid ways, or by helping people to rehearse a behavior. For instance, computer simulation may be used to provide interactive experiences that motivate and persuade physical activity and the substitution of walking for driving. Serious games, defined as digital games with non-entertainment purposes such as health care, security, management or learning [315], is another development in this area, particularly games for health, which focus on the impact games and game technologies can have on health care and policy.

For instance, medical pervasive gaming concepts are used to develop a spatial diary to assist diabetic patients and doctors to visualize deeply inscribed behavior, such as physical inactivity, stress or inadequate nutrition. These are factors that can negatively impact the patient's sugar levels. One prototype application is DiabetesCity—Collect Your Data, where young diabetics equipped with a standard blood glucose monitoring system, GPS-based positioning and mobile-phone based camera are playfully persuaded to document as much as possible of their medical data in their everyday life. Data from these devices can be imported to an Internet-based mapping tool (such as the browser-based Google maps or stand-alone Google Earth program) and shared with medical health professionals, enabling discussions of therapy and behavioral improvements on a visual basis by combining the medical data with the daily activities and their environments.

Finally, as a social actor, ICT can be persuasive by rewarding people with positive feedback, providing psychological influence by supporting feelings and providing empathy, as well as by taking on a social aspect by supporting various roles such as that of a doctor, nurse, rehabilitator, teacher, coach or guide. In [316], a smartphone-based physical health recommender system is evaluated which provides personalized and contextualized advice on physical activities, and just-in-time motivating advice at an appropriate time and location after taking environment, weather, user location and agenda into account. Participants noted that what they liked most were the timely reminders of daily activities that they tend to forget but are easy to do, such as walking to the coffee corner for a break, or a short lunch walk.

Ubiquitous Health and Wellness Monitoring: One result of the aforementioned technologies has been in ubiquitous healthcare due to environmental systems, which, in addition to using context-aware technologies that infer a subject's present state, make use of body-wearable sensors that collect, store, and share relevant data.

Developments in wearable sensors that have diagnostic as well as monitoring applications allow physiological and biochemical sensing, as well as motion sensing [71]. A new generation of wearable biometric sensors measuring Electrodermal Activity (also known as skin conductance or galvanic skin response) for monitoring arousal associated with emotion, cognition, and attention and mobile Photoplethysmography (PPG) for screening of cardio-vascular pathologies may find increasing safety, security and travel quality applications in transportation [317]. These sensor systems have applications in remote health telemonitoring not only in the home or community but also during travel or other everyday activities. Examples include BikeNet [312], Live City [318] and others.

Another trend within the ubiquitous health and well-being area is in in-vehicle biometric sensors, coupled with other sensors described in Chap. 2, to monitor the driver's psycho-physical condition and to obtain a holistic view of the driving situation, towards the goal of health telemonitoring and safety. As in other cases, these technologies have evolved over time and for a wide range of purposes including drowsiness detection and passenger weight and size detection for airbags. Biometric feedback may be obtained from drivers through sensors in the steering wheel, seat and seatbelt in order to monitor respiration, heart rate, sweaty palms and other pieces of data to assess the stress level of the driver.

Current research and prototypes have focused on biometric feedback to detect driver state, for example, stress, fatigue and inattention [319] and to detect driver intent, for example, lane-change intent [320]. The latter is achieved in a prototype system using Autonomous Cruise Control radar, Side Warning Assist radar, Lane Departure Warning Camera and a monocular head camera system to monitor the driver's head position and orientation. Aside from the in-car sensors described earlier, in an example of car-integrated medical sensor system, three systems were integrated into car seats for non-contact monitoring of vital signs, which omit the need for electro-conductive contact to the human body [321]. The three systems are capacitive electrocardiogram (cECG) monitoring, mechanical heart activity analysis (ballistocardiogram or BCG) and inductive impedance monitoring. Although the early results were somewhat mixed for the BCG signal, detecting such critical conditions in-vehicle could lead to the initiation of appropriate measures. Such measures could include in-vehicle driving interventions such as braking systems or safety auto pilots as a part of collision avoidance systems, V2V safety communications or automated contact with emergency management services.

3.4.3 Technologies to Meet Special Mobility Needs

As introduced in Chap. 1, ICT to meet mobility needs for persons with physical or functional disabilities encompass several areas of research, development and practice. An extensive literature has noted that peoples' right to independent living can be curtailed by transportation limitations [322, 323], thus limiting their ability to live independently. Universal Design (UD) has served as a paradigm for inclusion for persons with disabilities includes several concepts including barrier-free design or the retrofitting of buildings, facilities and other aspects of the built environment for all persons. It also includes accessible design for equal opportunity of access to mobility, facilities, devices, and services. Further, assistive technologies to enable people with disabilities to perform tasks which they previously could not by enhancing physical, sensory, and cognitive abilities fall within its purview, as does inclusive design of products and services [324].

Paratransit Service Management: One area of interest has been in the use of ICT to effectively manage paratransit services. Paratransit services are demand-responsive transit services (typically vans or minibuses) which have the ability to provide shared-ride, door-to-door or feeder services to persons with disabilities. The provision of paratransit service requires route planning and scheduling customer pick-up and drop-off times on the basis of received requests. Problems posed by multiple vehicles with limited capacity and temporal constraints (time windows) need to be considered. The problem of working out optimal service paths and times is called the Dial-a-Ride Problem (DaRP). More details may be found in [325]. Such transportation services are expensive to provide and, therefore, software that allows real-time scheduling, dispatching and coordinating of services and dynamic assignment of vehicles and routing is desirable.

Mobility Management: Such services are part of overall mobility management programs for persons with mobility needs and include notification to users of arrivals and delays, and online or phone-based trip reservations as in an airline system. These services may also protect a person's connections between transit vehicles at transit facilities and transfer points, and provide safety via sensors such as on-board cameras and collision detection, panic buttons, transit facility cameras, and automated activation of information and lights. Travel training, whereby car-dependent seniors or others with disabilities are given the skills to use regular public transportation services and community-based transportation services such as volunteer driver services, where drivers in the neighborhood volunteer their time and vehicles to give rides to seniors and persons with disabilities, may benefit from ICT use with respect to managing time and coordinating services, tracking driver reputation and incentives, and by otherwise connecting such programs to overall urban community engagement.

Wayfinding, Navigation and Outdoor Mobility: Location-based technologies have been widely used to assist wayfinding and navigation by blind and visually impaired (VI) persons in outdoor real-world environments. Technological solutions that aid in the orientation and mobility of VI persons can be classified into two groups: obstacle detectors and environment locaters. Obstacle detection was historically the purview of the white cane, seeing-eye dogs and human guides, whereas currently, a variety of Electronic Travel Aids, including proximity sensor-based contactless canes using mostly ultrasonic echolocation to prevent VI persons from walking into obstacles, are used. Electronic Orientation Aids assist in finding the way usually by providing VI persons with turn-by-turn directions.

Several items may be needed to augment a VI pedestrian's orientation in real-world environments, including location-sensing systems that have the high degree of accuracy necessary for pedestrian navigation, and map databases enriched with data layers reflecting the pedestrian network (sidewalks, pedestrian traffic control and crosswalks). Additional elements include non-visual auditory, somatosensory or olfactory landmarks (decision or confirmation points) that correspond to locations detectable by VI pedestrians such as pavement texture or inclination, odor and audio cues and points of interest (buildings, landmarks, retail) that can not only be travel destinations but also reference points that assist in creating a mental representation of the traveling environment. Additionally, mobile high performance computing devices, ubiquitous high-speed networking and high-quality non-visual (usually audio or tactile) user interfaces may be needed.

In one of several examples, [326] describes a wearable navigation assistive system consisting of positioning capability by a GPS receiver combined with a wearable inertial measurement unit (IMU, including accelerometers, an electronic compass and a pedometer), stereo cameras, and an enhanced GIS pedestrian-focused database. Examples of other ICT-based orientation and navigation aids include a system that uses on-bus GPS devices to send prompts to mobile devices carried by cognitively impaired people to remind them to get off buses at the right stop [327], and those augmenting the ability to complete complex travel tasks such as shopping using ordinary barcodes of products in supermarkets in contrast to RFID tags, which are typically used in more high-end merchandise [328].

Smart wheelchairs, walkers and even robotic scooters that autonomously follow a senior person home from shopping while transporting their purchased goods [329] are examples of mobility devices augmented by ICT. Augmented Reality (AR) (which augment the real environment with virtual computer-generated objects, thereby improving the quantity and quality of the information in the environment) and Spatially Augmented Reality (SAR) (which projects computer-generated surfaces into the real environment [330]) are increasingly playing an important role in assistive devices for mobility. In order to augment an environment, the important elements contained therein must be tracked using computer vision algorithms to locate known markers using data from webcams, whereas for complex tasks it may be necessary to implement inertial, magnetic or GPS tracking.

Indoor Mobility: Autonomous indoor navigation that facilitates mobility within the home or in unfamiliar buildings is a key aspect for independence. Most current location systems in GPS-denied indoor locations are not suitable for use by VI users since such systems extensively use visual components, which are not appropriate for VI users. In [331], an indoor VI navigation system is proposed based on 3D building models, while in [332], RFID is used to localize VI users inside a building. In [333], Bluetooth beacons and a map of the building are used to augment knowledge of the user's position. In addition to VI users, current approaches have addressed the specific autonomous indoor mobility needs of persons with other types of disabilities. One example is a Bluetooth trigger-based system for just-in-time turn-by-turn picture or video-based prompts for indoor and outdoor guidance to persons with cognitive disabilities such as traumatic brain injury, cerebral palsy, intellectual and developmental disabilities, schizophrenia, Down's syndrome, and Alzheimer's disease [334].

Another direction has been in the creation of whole-scale living and working environments that may have applications in airport terminals, public transportation transfer points, shopping malls and medical establishments. Ambient Assisted Living proposes systems where ICT is used to augment the independent daily activities of seniors and persons with disabilities in the community. A 2005 European Union report identified several technology options to be of high priority in developing AAL's: (1) microelectronics and microsystems, (2) embedded systems, (3) energy generation and control technologies, (4) new materials, (5) human machine interfaces, (6) communication, and (7) software, web and network technologies [335]. An early example is AmbienNet [336] which consists of an indoor location system based on sonar and radio frequency devices, smart wheelchairs acting as a mobile platform, with range sensors and embedded processors for guidance control, a vision based sensor network to collect and efficiently transmit context information (by means of a ZigBee wireless network), and a middleware layer implemented in order to ensure seamless communication and interoperability among the previously mentioned modules.

Chapter 4
Institutional and Policy Factors in ICT-Based Mobility Services

4.1 Introduction

In the two previous chapters, we reviewed developments in the areas of sensor technologies for mobility and examples of the mobility services they support. In this chapter, we discuss institutional, coordination and design strategies by means of which these technologies may be able to support sustainable mobility.

4.2 Institutional Challenges

Several institutional challenges arise with the start-up, growth and sustainability of the DMII over time. These challenges include the availability of adequate funding and revenue streams, the policy and regulatory environment within which the DMII operates, overall policies towards data, organizational and governance structure, and legal and ethical concerns.

4.2.1 Funding and Revenues

Funding the DMII means funding the hardware and software which collects and manages mobility information as well as the operations and services that are built on top of these assets. A wide range of stakeholders are involved in this activity, including government, which pays for the DMII through tax funds; private companies, which charge for the products and services they offer and even non-profit citizen hackers who contribute by connecting citizens to mobility information.

P. (Vonu) Thakuriah and D. G. Geers, *Transportation and Information*, SpringerBriefs in Computer Science, DOI: 10.1007/978-1-4614-7129-5_4, © The Author(s) 2013

4.2.1.1 Governmental Funding and Trends

Transportation sensing programs for a wide range of operations and management functions have been funded primarily by governments.

Governmental Support for ITS: Many countries have central strategic plans to support ITS deployment, and to varying degrees these plans have been implemented by means of real projects over the past few decades. In countries with a federal system of government, local transportation sensor programs such as those for toll collection, bus fleet management or smart card systems have been funded out of a combination of federal and local or regional transportation programs.

One challenge that has been noted is the lack of continued and consistent funding for pervasive transportation sensor systems, although this trend tends to vary widely across countries, modes and type of application. Several factors have been noted to contribute to governmental funding challenges, including lack of a consistent governmental programmatic framework that ensures funding at necessary levels, perceived high technology risks and perceived uncertain return on investments. For example, [337] notes that the US lags global leaders in ITS deployment, particularly Japan, Singapore, and South Korea, due to, among other reasons, a continued lack of adequate funding for ITS. The recent economic recession has also impacted the availability of governmental programs to a certain degree, although economic stimulus spending has thwarted some of the negative impacts.

Another challenge is politically designated government-funded ICT-based mobility projects. There is no doubt that project champions are needed to provide leadership regarding large-scale infrastructure including government-funded ICT projects. Earmarking resources over time for specific projects can ensure continuity and lead to desirable outcomes, in addition to ensuring that regions get a share of the total resources for transportation sensing projects. But it can also lead to implementation of projects that are the result of political preferences and not benchmarked on merits such as benefits, costs, opportunities and risks. Additionally, regions represented by stronger political forces may accumulate a greater share for projects that may not rank high on technical merit, thereby putting innovative projects at risk of underfunding.

Funding for Broadband and other Infrastructure: Virtually all services discussed here hinge on positioning systems such as GPS. GPS is funded primarily by the US Department of Defense with additional civil applications funded by the US Department of Transportation. The total military and civil funding for the GPS program was $1.5 billion in 2012 [338]. In the future, the EU's Galileo program, Russia's GLONASS program and other satellite navigation systems will play an important role in civilian location-aware applications.

Although telecommunications services in the developed world and in many developing countries are provided by the private sector with governments providing a regulatory, reform and oversight role, there has been an expanding public role in furthering the telecommunications infrastructure needed for the DMII. Recent governmental initiatives in the broadband infrastructure in Europe, US and other nations aim to expand wired and wireless broadband access. Based on the Digital Agenda for Europe, a flagship initiative of the European growth strategy (Europe 2020), the

EU plans to have 100 % broadband coverage by 2013 and 50 % or more of European households subscribing to Internet connections above 100 Mbps by 2020 [339]. Similarly, the National Broadband Network, which is currently being rolled-out in Australia, aims to provide fiber connections at up to 1 Gbps to 93 % of households by 2021 [340].

The US National Broadband Plan [341] developed with the American Recovery and Reinvestment Act (2009) similarly lays out an ambitious strategy for making broadband widely available with potential applications for smart homes, smart grid and smart transportation. These initiatives call for cost efficiencies over a wide range of efforts, ranging from coordination among civil works regarding digging and trenching, to streamlining laws and regulations concerning civil works, town planning, environment, public health and general administration to expedite broadband deployment. In the European case, the European Investment Bank (EIB) is currently lending an average of €2 billion each year to broadband projects; other sources include rural development and structural funds and numerous country-specific programs [342].

Public-Private Partnerships: To a limited extent, private participation in ICT-based transportation projects, particularly for toll road operations, has occurred through Public-Private Partnerships (PPP) or private contracting operations. Private participation especially relationships with technology partners were also leveraged successfully through PPPs in several Field Operational Tests (FOTs) of prototype systems. Early examples were the ADVANCE probe vehicle project [343] and the National Automated Highway System Consortium (NAHSC) in the US [344].

These demonstrations helped to highlight the institutional, contractual and project development challenges associated with PPPs along with challenges posed by conflicting goals and expectations of stakeholders. Conflict-of-interest difficulties can also arise in co-mingling roles of technology development/promotion with technology evaluation. The NAHSC study also highlighted the negative role that early-stage technology optimism can play with implementing novel transportation solutions in society and with the fact that while there may be no "technical showstoppers" [344], underestimating legal, institutional and societal challenges pertaining to public acceptance can be just as important in designing and building large-scale real-world transportation technology systems. Recent PPPs of transportation technology FOTs include the Mobile Millennium, a traffic-monitoring project using crowdsourcing in California [345], while the Michigan Connected Vehicle Testbed is a real-world operational test bed that offers testing and certification services for both the public and private sector [346]. Many of these FOTs involved university and R&D partners, in addition to public and private partners.

Among large-scale ITS deployments, the Vehicle Information and Communication System (VICS) in Japan, which provides information nationwide to in-vehicle navigation systems, is one of the largest operational PPPs [347, 348]. There are several worldwide examples of PPPs in operating Transportation Management Centers (TMC) where infrastructure costs are typically borne by governments and operating costs are split among partners. While this model has been successful, problems have been noted as well, including the inability of the private partner involved in the TMC to generate business out of supplemental, value-added traffic data to which they have access and consequent inability to carry on the for-free services to travelers which were a part of the PPP deal [349]. PPPs also arise in cases where a private company collects data for a government agency by using its own sensors including probe vehicles and other sensors [350].

4.2.1.2 Private Investments

Large amounts of wired and wireless communications infrastructure worldwide on which mobility services are based are funded privately.

Investments in Hardware and Communications Infrastructure: In its 2009 annual report, AT&T, the largest telecom provider in the US, noted that it invested more than $55 billion in communications infrastructure over the previous three years [351]. Other companies spent significant amounts on capital projects to increase network capacity. The majority of capital spending by large telecommunications companies is for purchasing software licenses, putting up cell sites, and new construction of wired networks [352].

Data centers and information technology assets used to produce mobile services including hardware and software are other areas where the private sector has invested significantly in order to enable ICT services. Google reported its 2009 information technology assets at US$3.9 billion [353]. Although only portions of these would carry communications traffic relating to mobility and location-aware requests, this load is expected to increase in the future, requiring greater private investments.

The World Bank notes that "private-sector led growth has revolutionized access to telecommunications services around the world over the past ten years, with every region of the developing world benefiting in terms of investment and rollout" [354]. The report also notes a recent estimate which suggests that in the developing world as a whole, the 2005 to 2010 investment requirements for new telecommunications capacity would have exceeded US$100 billion.

Revenue Streams for Software: The DMII is not only about hardware, sensors and communication systems but also software that captures mobility data and creates information of value. Revenue streams for such stakeholders are an important ingredient for the types of information that is generated and in their sustainability over time. Within the space of software and services, the LBS industry, for example, plays an important role by augmenting the DMII with mobility information that is collected by mobile sensing and information sharing services, location-based social networking and mobile commerce. The LBS industry is highly heterogeneous, with many different types of players. These include mobile operators, content providers, content aggregators, wireless application service providers, as well as others such as device vendors and manufacturers who are not directly part of the value chain of LBS provision but who play an important role because the technical characteristics of the mobile devices which they manufacture or otherwise support may enable, or impede the deployment of certain location-based services [355].

Business models for LBS include Business-to-Customer (B2C), Business-to-Business (B2B) and monetization of IP [356]. The authors of [356] also identify several revenue models for B2C commerce. First, the mobile service may be free. "Freebie" applications may be available to customers to entice early adopters and to build a user community with the aim of expanding the user base enough to introduce fee-for-service, or in the hope for a buy-out by a larger corporation. Second, they may be supported by advertising, such as fixed format banner advertising or mobile and location-based advertising. Third, mobile services may be available for a fee, which could be a one-time fee at the time an app is downloaded, when the service is first accessed, or on a subscription basis. In some cases, payment is required to access specific features within the service. This is the micropayment method [356].

In terms of B2B commerce, LBS revenue models may stem from advertising and charges from secondary use of user profiles, or transactions and location-based data collected by the mobile service. B2B commerce may also result from licensing fees for using Intellectual Property (IP) associated with the application [356].

Finally, revenue models may stem from licensing the mobile service application to third parties, usually in the form of a technology license agreement, to use, modify, or resell that property in exchange for compensation [356]. Compensation may take several forms including a lump sum royalty payment, royalties based on number of units sold or used or exchanged for the right to use a licensee's technology ("cross-licensing").

There is also a growing community of small independent "app" developers, some of which are non-profits and which may consist of a single person who works part-time on the development. Indeed some programmers build apps purely as a hobby. In such cases there may be no explicit revenue generation or business model since the reasons for production may stem from the altruistic aims of community service and the encouragement of civic participation through citizen engagement and community access to data (know as digital civic entrepreneurship activities). The apps or other products do, however, compete in mobile app marketplaces with products developed by larger companies and they will no doubt, play an increasingly important role in linking mobility services to users in the future.

4.2.2 Policy and Regulatory Environment

Transportation programs generally operate under a strict regulatory environment, given their safety, environmental, commercial and communications impacts. Of current importance in the context of mobility sensing are architectures for the deployment of ITS, policies on locational privacy, management of spectrum and emerging policy directions on distracted driving.

ITS Architectures: The purpose of ITS architectures is to guide system interoperability and information sharing among different components of an ITS and to "provide a framework for planning, defining, deploying, and integrating intelligent transportation systems" [357]. The architecture process can lead to many benefits.

For example, [357] notes the following as benefits: promoting the inclusion of desirable characteristics in the system, assistance in the creation of a national or regional vision for ITS, identifying and characterizing the major components of a country's ITS planning and in the interfaces between these components, and defining a framework into which future development and expansion activities can be targeted.

Well-known examples are the US National ITS Architecture and the European FRAME (FRamework Architecture Made for Europe). The use of the US architecture by US states wishing to deploy ITS is obligatory if federal financial support for those projects is desired, even if it is more costly to do so and even if local agencies see themselves as being constrained in their deployment strategies [358]. Some institutional challenges faced are given in [359]. The challenge noted for Europe is that there are many different states with different needs, making it impossible to develop a universal ITS architecture suitable for all of them [358]. For this reason, there is framework architecture and currently a number of European national and regional authorities have their own national and regional ITS architectures, based on the FRAME Architecture.

Locational Privacy: Sect. 2.6 described the major approaches taken to safeguard locational privacy from the technical point of view. Governmental approaches relating to locational privacy primarily address ways in which the consumer can be informed about potential privacy risks and provide the ability to opt-out or for consumer choice to be otherwise meaningful, so that personal data can be protected or used by third parties in ways that ensure personal privacy.

In general, US privacy policies tends to be fairly disaggregated, and a number of US laws such as the Privacy Act of 1974 are broadly applicable for many of the mobility services described here. Major privacy principles that are currently applicable to LBS providers are the Federal Trade Commission (FTC)'s "Fair Information Practice Principles" [360]: (1) Notice/Awareness whereby notice should be given to consumers before any personal information is collected in order for them to have the foundation for making an informed choice about whether or not to share; (2) Choice/Consent by means of which consumers are given the option not to share private data and for what it may be used; (3) Access/Participation, which addresses the ability of a consumer to access their personal data and correct, amend, or remove that data if it is incorrect; (4) Integrity/Security or the need for data collectors to ensure that data is stored and managed securely, both through technological and administrative means; (5) Enforcement/Redress, or providing mechanism to enforce the above principles. These, along with related guidelines from ITS America [361], the Vehicle Infrastructure Integration (VII) initiative (now Connected Vehicle) consortium and the CTIA—the Wireless Association [362], are applicable to mobility service developers and providers. Others, such as those underpinning trust services, for example, given by the American Institute of Certified Public Accountants, Inc., and Canadian Institute of Chartered Accountants (discussed further in Sect. 4.2.5), may also apply.

The EU Data Protection Directive [363], on the other hand, applies to any operation or set of operations which are performed upon personal data, called "processing" of data. The EU Directive stipulates that data processing must be fair and lawful and that data must be relevant and not excessive for the stated purposes. The purpose must be legitimate and made explicit. Data must be accurate and timely, and data subjects must be able to rectify, remove, or block data that is inaccurate. Additionally, EU member states must identify supervisory authorities to monitor the implementation of the Directive. Data on individual subjects must not be kept longer than necessary and used only for the stated purposes.

The approach taken by private companies also varies. Technical solutions to privacy have been adopted in varying degrees by mobility service providers but the major approach taken is notification by means of consumer privacy policies, which are either stand-alone documents or a part of overall Terms of Use of the Service.

Recent analysis showed that adoption of privacy policies appears to be fairly standard across the location industry [364]. By examining policies from 48 LBS, however, the authors of [364] found neither consistency nor completeness in how such policies are presented to the consumer. The reading level, as measured by the Flesch-Kincaid Grade Level, is roughly that of a sophomore to a junior in college, well above that of the average US adult. There is also limited or no information presented on how users may ensure accuracy and completeness of data, or steps to be taken in case of inaccuracies. Additionally, the lack of consideration or consistency regarding children's privacy is concerning, particularly as younger persons begin to use location and mobile services. Survey research of 382 LBS users indicated that less than half read the LBS policies to begin with [365]. The US Federal Communications Commission (FCC) recently reported that privacy should be considered by LBS providers at the earliest stages of product design [366]. This report stated that clarification is needed regarding the obligations, duties and responsibilities of LBS data generators and users to secure such data from unauthorized disclosure or access. In terms of timing and notification, LBS consumers may in some cases need to be notified of their privacy protection at several points in the use of a mobile service and there should not always be a feature to opt out of these messages.

Regulating and Managing Spectrum: The importance of having dedicated wireless spectrum for improving vehicle safety in cooperative transportation systems has been recognized by many governments. In 2004, the US FCC adopted licensing and service rules for the Dedicated Short Range Communications Service (DSRC) of the ITS Radio Service in the 5.850–5.925 GHz band (5.9 GHz band) [367]. Both public safety and non-public safety use of the 5.9 GHz band is licensed where DSRC involves both "safety of life communication transmitted from any vehicle, e.g., vehicle-to-vehicle imminent crash warnings, as well as communication transmitted by public safety entities, e.g., infrastructure-to-vehicle intersection collision warnings" [367]. In Europe, the harmonized use of radio spectrum in the 5875–5905 MHz frequency band for transportation safety-related applications has been specified [368]. The Australian government is considering a similar allocation of radio spectrum at 5.9 GHz for ITS [369] and new licences are embargoed.

There are several institutional issues related to deployment of DSRC in vehicles [369, 370]. It will be necessary to preserve the 5.9 GHz DSRC spectrum for cooperative transportation systems and to ensure that the integrity of the network is maintained and prioritized for safety critical applications by commercial organizations. There is also a need to establish processes to administer the 5.9 GHz DSRC certification and compliance program for in-vehicle units and site specific licensing of roadside equipment. Standards will be needed, as will regulations regarding information security and privacy.

Distracted Driving: Distracted driving has been the subject of various policies, of which the ban on using cell phones (talking or texting) while driving is the most common. More comprehensive policies address a larger number of factors related to distracted driving such as telematics design and driver education. Various stakeholders are involved in the process, including governmental organizations, industry associations, international standards-making bodies and the automotive and telematics industries.

Voluntary Phase 1 guidelines were proposed by the US National Highway Traffic Safety Administration which "apply to communications, entertainment, information gathering, and navigation devices or functions that are not required to operate the vehicle safely and that are operated by the driver through visual-manual means (meaning the driver looking at a device, manipulating a device-related control with the driver's hand, and watching for visual feedback)" [371]. These guidelines include recommendations to: (1) Reduce complexity and task length required by the device; (2) Limit device operation to one hand only (leaving the other hand to remain on the steering wheel to control the vehicle); (3) Limit individual off-road glances required for device operation to no more than two seconds in duration; (4) Limit unnecessary visual information in the driver's field of view; and (5) Limit the amount of manual input required for device operation.

Additional guidelines recommend that devices providing dynamic (i.e., moving) non safety-related visual information should provide a means by which that information is not seen by a driver. Phase II of the proposed guidelines may address devices or systems that are not built into the vehicle but into aftermarket and portable mobile devices. Phase III may involve guidelines relating to voice-activated controls to further minimize distraction in factory-installed, aftermarket, and portable devices. Recent surface transportation legislation in the US, the Moving Ahead for Progress in the 21st Century Act (MAP-21), has placed emphasis on several aspects of distracted driving, including consistent governmental funding for incentive-based educational, testing and notification programs.

4.2.3 Policies Towards Data

We consider two recent policy trends regarding mobility data: Open Data policies stemming from open government and Big Data initiatives. These trends are likely to stimulate R&D activities relating to pervasive mobility sensing environments.

4.2.3.1 Open Data Policies

The DMII would benefit from administrative and planning data traditionally held by government agencies. Open data policies promote the idea that certain data should be accessible to everyone to use and republish without copyright or other restrictions. Governmental open data policies seek to create a knowledgeable, engaged, creative citizenry, and to foster innovations in mobility products and services. They also seek to bring about accountability and transparency. Smart cities are noted to require an innovative set of services that provide "information to all citizens and businesses about all aspects of city life via interactive, city-wide, Internet-based applications" [372]. Open data policies supported by the US Open Government Directive [373], Declaration of Open Government following the Government 2.0 Taskforce report in Australia [374], the European Commission communication on open data for "all the information that public bodies in the European Union produce, collect or pay for" [375], the UK Putting the Frontline First: Smarter Government Action Plan to "radically open up data and promote transparency" [376], and others have stimulated initiatives in open data from the highest levels.

The vision of the DMII as a platform to seed mobility applications and planning decision support partially rests on the accessibility of various data streams. Being able to merge mobility data with other non-traditional data such as crime, socio-demographics, built environment or housing may lead to mobility intelligence which may otherwise not be possible. However, open data initiatives can face many challenges. In a study of open data policies in five countries, the following are noted as the top 10 barriers: closed government culture, privacy legislation, limitations in data quality which prohibits it from being published, limited user-friendliness, lack of standardization of open data policies, perceived security threats which prohibit certain types of data from being published, uncertain economic impacts which has led some countries to question the value of open data policies, questions of the digital divide which presumably may create inequalities in access, and budget cuts and consequent limited capacity of existing networks [377].

Additionally [377] noted that some countries identified existing charging models as barriers because the income is derived by some government organizations from the sale of data, which makes them reluctant to freely publish their data. The latter may be an important item for discussion in the transportation sector since government-owned and funded sensing infrastructure which generates pervasive mobility data can be an income-generating asset which has been limited to date. Many cities around the world now have data portals where government agencies upload digital information that is license-free and in non-proprietary formats.

Recent trends in government-supported APIs and mashups have enabled the ability to tap into such data in a growing number of ways. There are three groups of users in this space which have used such data for mobility services: technology/ICT companies who tap into such data for commercial purposes, ICT-oriented social entrepreneurs (citizen hackers) who produce curated information including mashups that convey relevant events and apps for others to use ("apps for democracy") and ordinary citizens who access the data to understand more about transportation and other conditions in their communities.

Public transit agencies have been active in making their data available to developers through APIs and these developers in turn have created a variety of applications relating to transit, the most common of which are real-time transit arrival systems, and transit-based mobile and web-based trip planning applications. Other transportation agencies make data available through Memoranda of Understanding and the like, and in some cases there may be exclusivity factors involved with respect to data-sharing between an agency and a specific private company. However, it is possible that the new wave of open data policies and the trends set by many forward-thinking agencies may render such arrangements obsolete. More research is needed into how revenue models for government-funded data can be derived in the era of open government data.

4.2.3.2 Big Data Initiatives

"Big Data" is the term being applied to very large volumes of data which are difficult to handle using traditional data management and analysis methods. This topic has generated a lot of recent interest. Such data arises in the mobility sector from pervasive sensing, open government sources, transactions (in transit, toll systems and LBS), and business and operations management and scientific research. Social media information sources, and the highly unstructured data they yield (text, images, video) are Big Data aspects that are also likely to be useful for mobility intelligence. This interest is well-aligned with the overall excitement of the potential for extracting value-added intelligence and deriving insights.

Synergistic developments in Data Science or Big Data analytics have stimulated scientific and curricular developments in computer science, statistics and the geospatial and GIScience disciplines. Developments in mobility mining, spatial knowledge discovery, geographic data mining and related areas will be aspects of Data Science needed to analyze Big Data in the mobility sector. The White House announced an investment of over US$200 million in Big Data projects in six federal agencies (not including the Department of Transportation) under the "Big Data Research and Development Initiative" [378]. The emphasis is on fundamental research needed to advance state-of-the-art core technologies needed to collect, store, preserve, manage, analyze, and share Big Data.

An additional focus of the Big Data initiative is on teaching and workforce development in Big Data technologies. Such governmental initiatives may create the state of readiness needed to elicit benefits from mobility Big Data. A shortage of the analytical and managerial talent necessary to make the most of Big Data has been noted to be significant, and the United States alone faces a shortage of 140,000–190,000 people with deep analytical skills, as well as 1.5 million managers and analysts to analyze Big Data and to make decisions based on their findings [379].

4.2.4 Management and Governance

The public management of the primary tier of the DMII is in the hands of multiple agencies at different levels of government. We focus on capacity building and management strategies needed to support DMII activities.

4.2.4.1 Capacity Building and Body of Knowledge for Mobility Intelligence

The discussion in the previous section suggests that, overall, the successful utilization of the DMII would require several innovations in capacity building and workforce development. A Body of Knowledge (BOK) typically describes the knowledge of a field in an organized way and "consists of comprehensive inventory of the intellectual content that defines a field" [380]. The ACM/IEEE-CS Joint Task Force on Computing Curricula 2001 and Geographic Information Science and Technology (GI&T) Body of Knowledge are well-known examples.

The main knowledge areas within the BOK for mobility intelligence would include: (1) broad-based understanding of multi-modal transportation systems particularly in relation to other urban systems, sustainable urban development and planning; (2) exposure to the conceptual and multidisciplinary foundations of understanding mobility and ICT, and traveler behavior and use in relation to ICT; (3) methods for information management including knowledge of communications, sensor technologies, sensor fusion, methods for information extraction, moving objects database management, human-computer interactions, intelligent systems, mobile applications and information security; (4) multidisciplinary methods for the analysis of mobility information drawing upon methods on travel forecasting, prediction and operations management from transportation engineering, planning, geography, as well as from the computational disciplines especially from areas such as data mining and knowledge discovery; (5) ICT-based mobility program development and management such as procurement, contracting, intellectual property, market analysis and related areas; and (6) policy analysis and evaluation of broader economic, social and behavioral impacts; and (7) legal and ethical issues relating to the pervasive sensing environment, for example, privacy, information security and liability. These seven elements would round off the BOK that would be necessary to establish a professional domain that deals with the DMII.

One example of a broad-based capacity-building approach at the level of doctoral research is the interdisciplinary Computational Transportation Science program funded by the US National Science Foundation's Integrative Graduate Education and Research Traineeship (IGERT) Program [381, 382]. Participating departments in this doctoral program in the University of Illinois at Chicago, which recently ended, were Computer Science, Urban Planning, Civil Engineering, Information and Decision Sciences and Law. The required coursework consisted of urban transportation planning and policy, urban travel demand forecasting, information technology and economics of urban transportation, principles of computational transportation science including moving objects database management, sensing and communications technologies, and legal issues relating to locational privacy, security and ethics. The majority of the dissertations focused on computational, mobility service development, policy analysis and management issues around mobility intelligence.

While the content of the areas covered in this book may change over time, the conceptual aspects of a broad-based BOK that draw from relevant areas of urban planning, civil engineering, computer science, geography, economics, management, sociology and law, among others, will be important in facilitating such capacity building efforts.

4.2.4.2 Governance and Public Management

ICT-based governance is a collection of technologies, people, policies, practices, resources, social norms and information that interact to support governing activities relating to mobility management [383].

ICT-based Mobility Project Leadership: Ingredients identified as being important for successful public ICT and e-governance activities are the presence of a champion who will collaborate with stakeholders and provide leadership [384], has social and technical skills, including team building and project management skills [385]. Experience with ITS early deployment studies discussed in Sect. 4.2.1.1 indicated that knowledge of project management where design, research and implementation occurs concurrently and complex contractual knowledge with the involvement of multiple partners are experiences called for. Other factors include the ability to manage expectations among stakeholders with conflicting goals and to deal with conflict-of-interest difficulties which arise by having partners who promote projects and also lead their evaluation.

ICT-based Mobility Project Development and Management: One novel aspect of governing the mobility sensing enterprise is involving people in mobility intelligence. Such a strategy may involve public bodies, the private sector and ordinary citizens jointly, in the creation of information. Problems relating to formation, development, growth, management of data quality and other issues were discussed in Sect. 3.3.3. Legal and ethical issues are addressed in Sect. 4.2.5.

Involving citizens in transportation-related problem-solving, idea generation and visioning has a long history. However, the extent to which these activities ultimately translate to citizen power regarding important decisions about transportation projects and operations has been examined to a lesser degree, suggesting that some of these activities may equate to going through the empty ritual of participation without having the real power needed to meaningfully affect the outcome of the process. As early as 1969, [386] proposed the ladder of citizen participation with eight levels of participation ranging from manipulation, where citizens are placed in rubberstamp advisory committees or advisory boards by government agencies for the express purpose of educating them or engineering their support for projects which are already selected.

Addressing Equity in Service Provision: The need to address the digital divide and potential spatial and socio-demographic inequalities in access to mobility information is an issue that needs further work. People from all geographical areas and socio-demographic backgrounds may not be equally able, likely or interested in participating and co-producing. This can lead to inequalities in mobility service provision. More research is needed in these areas to examine the participation model and the extent to which it can be a sustained model for mobility intelligence. SMARTiP, a recent EU-funded 4P project, aims to identify innovative and sustainable ways of building the capacity of citizens and public services to work together with technology leaders and digital developers to co-produce future internet enabled services, including mobility services, on the basis of the "widest possible digital inclusion" [387]. Demonstration projects and FOTs would be needed to examine these issues in greater detail.

4.2.5 Legal and Ethical Issues

Legal considerations surround many issues covered in this book, ranging from LBS to UGC. Aside from the issues discussed under regulation, major issues are liability, intellectual property infringement and consumer awareness and protection.

Legal issues surrounding liability underlie much of the mobility services offered. Safety concerns relating to injuries and property damage are the primary drivers of liability concerns. The issue of who carries responsibility can be complex to determine with fast-moving technology developments. Liability can be incurred through a wide range of legal mechanisms, which vary between different jurisdictions. Product liability is liability for personal injury or property damage caused by use of mobility products and services. Under modern concepts of product liability, there are three theories an injured party can assert against the producer or seller of a defective product: negligence, breach of contract (breach of warranty) and strict tort liability [388].

For government-operated ITS, additional considerations relating to liability of public bodies may be evoked, which is complex area within the law. Courts have in many cases limited or denied the liability of public bodies. Establishing liability against a public body for a failure of infrastructure is comparatively rarely achieved [389] (for example, road authorities are not normally liable for damage caused by poor road surfacing, even though this can result in serious accidents). The authors note that methods of transportation that rely more on complex systems maintained by public bodies generally tend to see a higher rate of successful litigation in the event of a failure. They cite examples of failure of rail or air-travel related infrastructure.

The creation and use of UGC opens up interesting legal questions as well. UGC is different from the way rights to and liabilities arising from traditional content created by authors, artists and the like are allocated [390]. With such traditional content, creators and distributors enter into binding, bilaterally negotiated contracts that can include representations and warranties, indemnification clauses, and other familiar language allocating their respective liabilities and designating ownership rights. In the sphere of UGC, these issues are typically dealt with through End User License Agreements (EULAs). Despite such license agreements, however, copyright infringement issues continue to emerge from the use of UGC.

Based on [391], three doctrines of copyright infringement may be relevant to the case of UGC derived from mobility idea generation, crowdsourcing or design competition: direct infringement, contributory infringement, and vicarious infringement. To prevail under a theory of direct copyright infringement, the authors noted that a plaintiff must demonstrate it owns the copyright in the work and that one or more of the plaintiff's exclusive rights given by the Copyright Act was violated, for example, by reproducing the work or distributing copies of the work. Contributory infringement may be established by demonstrating an objective of promoting copyright infringement, for example, by the party having knowledge of the infringing activity or when "the party induces or materially contributes to the infringing conduct of a direct infringer". Vicarious infringement may be demonstrated if the party has "the right and ability to supervise the direct infringer", and has a "direct financial interest in the infringing activity" [391]. In addition to copyright issues and rights clearances, legal issues relating to offensive, defamatory or illegal content, minors and the likelihood of action by authorities may be of concern [392].

From the perspective of a mobility service provider which solicits content through crowdsourcing and idea competitions, protection may be available from copyright infringement claims under the Digital Millennium Copyright Act (DMCA), Section 512 of the DMCA, contains "safe-harbor" provisions for online service providers. These safe harbor provisions may limit the liability for copyright infringement as long as "notice-and-takedown" procedures are established, so that content is removed promptly when a copyright owner notifies the service provider of infringement, and the service provider has no actual or effective knowledge that the material in question is infringing.

There has been limited legal research on information extraction from UGC through web content mining, popularly known as web scraping. An example of web scraping in mobility intelligence is these of unstructured (text-based or image) information published in newspapers or submitted by users to blogs or microblogs about special events or other conditions such as road construction that could potentially affect traffic. Several ethical considerations are noted to apply to retroactive UGC [393], which, as described in Sects. 2.2.3 and 3.3.2, refers to content submitted by users for purposes other than that for which it is ultimately used or for which the user did not intend to contribute or was not aware of its ultimate use by the web mining activity. The authors of [393] note that because the producer of the UGC is not aware how the information they submitted is being used or that it is even being used, and was not given the opportunity to consent or withhold consent regarding collection and use, that this is ultimately an issue of threat to the individual data producer's privacy.

Additionally, the creation of user profiles through web content mining can lead to anonymous profiles and de-individualization [393, 394], thereby erasing direct connections, and if used for making policies, underlying values like "non-discrimination, fair judgment and fair treatment" [393] may be threatened. Of course, such groupings are very often the subject of transportation policy, but decision-making on the basis such groupings are often supported by additional data (from surveys or administrative data) that can nuance judgments.

Further, with respect to web content mining, from the perspective of a web resource owner, the Terms of Use document is used to notify users of copyright infringement policies and to list procedures for claimed copyright infringement with respect to UGC. The Terms of Use statement may also be used to notify web content miners of their "no-scraping" policies, usually with statements such as "You may not scrape or otherwise copy our Content without written permission". The extent to which these non-scraping policies are enforceable in a court remains a gray area with no clear legal standards.

Overall, with content that users submit, neutral third-party service organizations may be able to provide guarantees of data integrity, security and privacy to consumers. The term "trust services" as defined by the American Institute of Certified Public Accountants (AICPA) and the Canadian Institute of Chartered Accountants (CICA) are a set of professional attestation and advisory services based on a core set of principles and criteria that address the risks and opportunities of IT-enabled systems and privacy programs [395].

Trust Service principles and criteria are organized into four broad areas, primarily regarding privacy policies and Terms of Use of "entities" in IT, e-commerce, e-business, and systems: (1) *Policies*: The entity has defined and documented its policies relevant to the particular principle; (2) *Communications*: The entity has communicated its defined policies to authorized users; (3) *Procedures*: The entity uses procedures to achieve its objectives in accordance with its defined policies; (4) *Monitoring*: The entity monitors the system and takes action to maintain compliance with its defined policies.

Principles and criteria developed by the AICPA/CICA for use are: (1) *Security*: The system is protected against unauthorized access (both physical and logical); (2) *Availability*: The system is available for operation and use as committed or agreed; (3) *Processing Integrity*: System processing is complete, accurate, timely, and authorized; (4) *Confidentiality*: Information designated as confidential is protected as committed or agreed; (5) *Privacy*: Personal information is collected, used, retained, and disclosed in conformity with the commitments in the entity's privacy notice and with the 10 criteria set forth in an appendix on Generally Accepted Privacy Policy (GAPP) relating management, notice, choice and consent, collection, use and retention, access, disclosure to third parties, security for privacy, quality, and monitoring and enforcement [395].

There are currently several services that offer certificates or seals of approval. It has been noted that because LBS are still in an early stage of diffusion when potential consumers do not yet have credible, meaningful information or relationships with service providers, examining consumer trust is important, particularly as LBS providers need to engender sufficient trust to persuade first-time consumers to transact with them [396]. The results of these authors indicate that by joining third party privacy seal programs and introducing device-based privacy enhancing features, service providers could increase consumers' trust beliefs and mitigate their privacy risk perceptions. Third-party groups themselves need to earn the trust of consumers by clear and acceptable principles and a credible examination of audit practice [397].

4.3 Societal Preparedness

Given the rapidly changing nature of technological developments, it may be difficult to address consumer protection through technology design and regulation alone. Approaches to expedite the level of societal preparedness in terms of digital citizenship and by addressing the digital divide are also necessary elements.

4.3.1 Digital Citizenship

Awareness is needed among citizens regarding expected behaviors, rights and responsibilities and ways in which to exercise such rights and responsibilities. Digital citizenship has been defined as "the norms of behavior with regard to technology use" [398]. The literature on this topic is generally on K-12 education and Internet use and very little work has been done on digital citizenship in the mobile environment. These rights and responsibilities can be summarized under three areas of behavior: (1) digital etiquette regarding net and device (personal mobile device or vehicular) use; (2) digital literacy or education about transportation technology and their use; and (3) digital participation regarding deliberation, sensing and sharing in the mobile environment.

Digital Mobile Etiquette: Two aspects of etiquette are relevant in the digital mobility environment. (1) *Etiquette relating to device use*: Etiquette considerations arise from annoyance and travel quality consideration for others as well as that of safety. Voice communication is not desirable in public transportation spaces [399] and while the overall trends are likely to depend on societal norms and acceptance and it is possible that with accurate low-cost indoor positioning, cell phones may be able to go into silent mode without active intervention by users. Another aspect relates to safety concerns resulting from distracted driving or walking due to the use of mobile devices while on the move. In terms of interventions, legal and enforcement strategies have been the primary approaches to date to address distracted mobility (as discussed in Sect. 4.2.2). It would appear that avoiding distracted mobility should become a part of digital literacy skills and the focus of social marketing.

In terms of technology, there are a range of solutions already available in the market such as passive blocking of incoming calls and text messages while driving, as well as active driver warning systems. (2) *Etiquette relating to online behavior*: Etiquette considerations in the mobile sphere may arise from online communication and content submission. One cause of poor etiquette in online or digital communication is the absence of feedback and reduced sense of responsibility between people who may never have to address each other face-to-face [400]. This results in lack of clarity regarding where accountability and responsibility reside in virtual communities. The result of such ambiguity may be lowered accuracy of UGC, potentially leading to inaccurate or even dangerous information in some cases.

Digital Literacy: Digital literacy in the mobile information environment comprises several skills and competencies—however, it is "more than the mere ability to use software or operate a digital device; it includes a large variety of complex cognitive, motor, sociological, and emotional skills, which users need in order to function effectively in digital environments" [401]. Digital literacy in the mobile information environment may include the following elements: (1) *Information access and use*: One constituent element is basic skills and core competencies in locating, accessing, understanding, using and potentially sharing information related to transportation and mobility effectively and efficiently. (2) *Comprehension of implications of use*: Another element is understanding the implications of information access and use and constructing knowledge, for example, regarding potential risks to privacy and information security. (3) *Skills for protecting safety and security and valuation of benefits versus risks*: A third element is skills and competencies to protect oneself from such risks (for example, knowledge of preference settings, rights afforded by privacy policies and terms of use) and to make judgments regarding how the benefits afforded by the piece of information compares to risks. (4) *Comprehension of cyberspace norms*: Digital literacy also extends to having a realistic understanding of the "rules" that prevail in cyberspace, including the etiquette for responsible online reporting, as mentioned previously and other factors such as plagiarism and copyright infringement, termed reproduction literacy [401].

Digital participation: Various authors have noted that participation through ICT is an essential element of digital democracy and citizenship (for example, [402]). As noted in Chap. 3, citizens can actively participate in transportation decision-making in various ways but these information modes may not be fully realized and the evidence regarding the extent to which such participation occurs and the extent of their inclusiveness and sustainability over time are either mixed or non-existent. This could be due to many factors, such as high digital literacy levels required, lack of time, interest or information on how to participate or overwhelming amounts of information required to participate. Technological elites, organized interests and individuals with strong opinions and consumers with higher buying power and early adopters of technology may variously dominate early adoption of transportation technology, online public discourse and sensing modes. Finally, the extent to which ordinary citizens can truly make a difference in the public management of a complex sociotechnical system has not been highlighted in the published literature or the media. These barriers would need to be overcome in order for genuine digital participation to occur.

4.3.2 Digital Divide

Benefits from the increasing use of ICT in transportation may not be equally available even though technology has become cheaper and more abundant. Much has been written about the digital divide or "inequality of access to information technology and the Internet"—a dichotomous gulf between the technological "haves" and "have-nots" [403]. The digital divide may be present across different locations, socio-demographic groups and also across different stages of the same person's life. In the digital mobile environment, the meaning of this term is somewhat ambiguous. There are already strong differences in the quality and quantity of transportation services across different locations, socio-demographic groups and lifecycle stages. For the digital mobile environment to have geographical and social inequality consequences, lack of access and availability of digital technologies have to magnify these already-present negative mobility and accessibility effects.

A discussion of the mobile digital divide requires clarification of what, who and where the divide pertains to. Regarding what, mobile digital divide means inequality in access to the Internet, mobile devices and ICT-based fixed or vehicular transportation infrastructure and services. Regarding who, although lack of such access may happen due to several reasons, including technical reasons, the mobile digital divide has social inequity underpinnings. Inequalities may be income, gender, race, age or physical and functional ability-based. Finally, regarding where, there may be country-to-country differences, urban-rural differences and intra-urban differences.

To date, there is no study which has systematically examined these various aspects of the mobile digital divide. The effects are likely mixed since although ICT has brought about benefits to marginalized groups or areas in some ways, it has also contributed to an increase in the divide in other ways. For example, urban areas have generally benefitted from investments in intelligent transportation technologies due to their congestion-management potential, including by means of ICT deployments for urban public transportation which serves travel needs of lower-income residents. However, rural areas have not benefitted to the same degree, although ICT investments for flexible demand-responsive services may support mobility of transit-dependent rural residents [404]. Seniors make up only a small proportion of those online and therefore may not be able to participate in the digital mobile environment to the same extent as younger persons, however, vehicle and infrastructure-based technologies such as those discussed in Chap. 3 may benefit seniors. Finally, although mobile technologies have been enthusiastically heralded by some segments of industry as bringing an end to digital divide concerns, issues of cost, quality and digital literacy still remain barriers to equal participation in the digital mobile environment [405, 406].

4.4 Coordination of ICT with Transportation Services and Plans

It was hoped that with the use of ICT and improved information for sustainability and less congested options for travel, users would be motivated to switch to more sustainable modes of transportation, choose less congested routes or change trip departure times or to telecommute instead of traveling to work. Yet these trends have not been noticeable to date. ICT can, in some ways, reduce travel or lead to more sustainable and productive travel, but in other ways, stimulate more travel, which in the aggregate, has negative effects. A particular ICT may have more pronounced (positive or negative) travel outcomes than other ICT, leading to difficulties in overall generalizations of effects. The travel effects of the ICT may also evolve over time as a result of the change in the design of the ICT itself or due to differences in the level of societal preparedness, attitudes and the ability to substitute travel for ICT use. Further, the time or resources spent in travel may also be utilized differently as a result of changes in adoption and domestication of newer types of ICT.

A voluminous literature pertaining to telecommunication and travel is summarized in [407]. The author presents twelve reasons why the increased use of ICT may not lead to reduced travel. These twelve reasons are divided into two groups: passive reasons which help explain why ICT does not always automatically substitute for travel versus active reasons which describe mechanisms by which ICT actively increases travel. Among the passive reasons given, the need to physically transport humans and/or goods, lack of an ICT counterpart for such transportation activities, unavailability and inaccessibility of these ICT solutions even if they are available and the need for face-to-face interactions, the potential for social encounters, and the opportunity for breaks from being sedentary or isolated, are some examples. ICT can also directly stimulate new travel by inviting travel directly, by increasing accessibility to people and places for sharing and interaction, and by improving congestion and travel speeds.

Many of the environmental and social outcomes targeted by mobile travel technologies will be possible only if there are transportation alternatives that allow the user to undertake eco-friendly, healthy and sustainable travel. There is a need to strongly couple ICT-based mobility services with services and plans that support sustainable transportation options and to make ICT-based mobility strategies an integral part of infrastructure planning. In addition, technologies should follow certain design principles which may allow broader goals to be achieved more easily.

4.4.1 Coordination with Transportation Services

Mobility information services should be strongly coordinated with sustainable transportation choices. The area of Transportation Demand Management (TDM) addresses such strategies. TDM includes services to encourage alternatives to the single-occupant vehicle such as carpools, vanpools, transit, cycling, and walking. Compressed work weeks, flextime, telecommuting and other alternative work-related programs are also TDMs, as are congestion charging for driving in congested areas or during congested times and parking management tactics such as preferential parking for carpools and parking pricing.

Examples of ways in which TDM strategies can be supported by ICTs are as follows:

Public transportation: Real-time transit arrival systems, travel with the aid of transit itinerary and activity scheduling systems, supported by personalized mobility systems, transit-focused recommender systems for shopping and entertainment; collaborative and shared transportation for the "last-mile" (to and from transit stations); *Shared transportation and Mobility on Demand strategies*: Systems supporting access-based demand including real-time search, scheduling, reservation and payment (if necessary) for ride-sharing, car-sharing, bike-sharing and walking group formation;

Active transportation: Services to support opportunistic walking and cycling to meet daily fitness goals, information supporting intermodal connection by transit, shared transportation, walking or cycling, and routing and planning for safe travel;

Neighborhood mobility management: Ride coordination for seniors in volunteer driver systems, crowdsourced congestion or trouble spots in neighborhoods for safe walking or transit use, real-time walking buddy finders from transit stations and stops.

In many of the above examples, the presence of safe and reliable travel alternatives to the private car is assumed. This may not be true in all cases. In fact, historical transportation investment trends in some countries have supported car travel to the extent that the basic infrastructure necessary for non-car travel is limited. Low-density developments, precipitated by car dependence, in turn, do not support the public transportation ridership volumes necessary to support their existence and result in decentralized urban development of residential, work, shopping and other destinations such that shared transportation becomes difficult. These patterns may also act as barriers to businesses that offer shared transportation, such as car-sharing, bicycle-sharing and ICT-based collaborative travel opportunities.

Secondly, even if sustainable transportation options are present, it may be much more convenient for a person to drive alone due to savings in door-to-door travel time, difficulties in scheduling, mismatch in work destinations and schedules, need to make multiple stops for multiple activities and so on. Low fuel prices, low licensing and insurance costs, easy availability of auto loan credit and low parking costs may continue to support solo car travel in some cases.

Third, social and cultural values may support personal mobility afforded by the personal car, and the car is the socially desired and mainstream method of transportation in many areas across the world. In general, travel behavior suffers from hysteresis and changing from car use to say, public transportation or walking or cycling, is resisted due to prevailing habits, attitudes, preferences and subjective norms (a far from complete list of references in this area includes [408, 409, 410, 411]).

Nevertheless, markets currently exist in areas with such urban development forms where the land-use, population densities and jobs-housing-shopping balance can offer a role for ICT-based mobility for sustainable public, shared and active transportation. A study in the City of Chicago, for instance, [412] noted that the bus system experienced an overall increase in ridership after real-time bus arrival information became available, after statistically controlling for employment, gas prices and a number of other factors that also affect bus ridership. It should be noted that this study, which examined changes in aggregated ridership data, is limited by the fact that it did not make a distinction between whether the gains in ridership accrued from new users (for example, car users) who were attracted to the public transportation systems or due to existing public transportation users who used the bus service more frequently due to the convenience afforded by bus arrival time availability. Geographical disparities were noted to exist in the extent to which ridership changed after real-time bus arrival became available, with the greatest changes in bus boarding estimated to have occurred at bus stops surrounded by higher densities, plenty of jobs, shopping, medical, or entertainment opportunities and overall population characteristics that were less representative of the digital divide [413].

The nature of personalized mobility is itself slowly changing, with increasing emphasis on hybrid and electric vehicles for daily travel and neighborhood, ultra-subcompact eco-friendly vehicles that address mobility-on-demand and the public transportation "last-mile" problem [414]. Investments in electric vehicle charging stations and ultra-compact vehicle parking inside buildings and public areas may continue to revolutionize sustainable personal transportation modes. Increasing levels of automated vehicles in real-world traffic may improve safety by reducing human error although such potential remains to be empirically verified.

It also remains to be seen whether and to what extent the recent "cool" factor associated with emerging mobility information systems, social media and apps can transform societal values towards public, shared and active transportation modes. This may partly be a generational issue, and the link between ICT and mobility outcomes may be strengthened by younger persons, particularly Digital Natives [415] who were born after 1980 when social digital technologies came online, and who, due to greater exposure since a young age, use such technologies to a much greater degree for their everyday activities, compared to earlier generations. A similar cohort, Milliennials [416] who were aged 18–33 years in 2010 are noted to be more likely to access the Internet wirelessly with a laptop or mobile phone compared to older generations. In addition, they clearly surpass their elders online when it comes to the following factors: use of social networking sites, use of instant messaging, using online classifieds, listening to music, playing online games, reading blogs, and participating in virtual worlds.

Generational trends exist in automobile ownership and use as well. Three generations in US society were examined in order to examine how car ownership trends have changed among young adults (18–24 years of age) over the 40-year period, from the mid-1960s to the mid-2000s [417]. Individuals in later generations (who were young adults in the late 1990s and early 2000s) were found to be significantly more likely to enter the car market at an earlier age compared to earlier generations. Emerging research is showing, however, that the number of young people with drivers' licenses is down in the US [418]. It is tempting to think that this is the result of teens switching from cruising in cars to social media and gaming environments, but there are probably a number of other factors at work. For example, according to recent industry research, 60 % of US parents whose teenage children currently hold a license reported that the economic downturn at the time of writing this book led them to reduce saving for or spending on their child's driving, including the cost of a car and other related expenses [419].

Nevertheless, young adulthood plays an important role in the individual's later life and it is possible that some of the values of a driving-free youth may remain in the future although this can only be verified in the future with longitudinal data. But it is possible that such trends with young persons may continue, laying the groundwork for a stronger role for ICT-based mobility services to contribute to sustainable travel outcomes in the future.

While the conditions necessary for ICT-based mobility strategies to strongly support certain forms of sustainable outcomes is either not extensive, or is close to non-existent in some cases, there is hope for the future given ongoing planning and social trends. A multi-pronged approach that includes urban development planning and educational and social development strategies will be needed to address the broad spectrum of factors that currently act as barriers.

4.4.2 Coordination with Transportation Planning Efforts

In addition to coordinating ICT-based mobility strategies with the design of TDMs and long-term transportation strategies, challenges remain in connecting the use of data from the primary tier of the DMII to long-range, regional transportation planning models. These models are technical simulations to forecast demand for transportation infrastructure and services in future years (10–40 years). The mainstay of these efforts is the four-step modeling process that had their origin in the 1960s. The results from these models are an important part of the information-base that guide a region's investments in transportation, sometimes over many decades. A central source of data used in this process are household travel surveys, which gather detailed information on households' socio-demographic and travel patterns using Computer-Assisted Telephone Interviews (CATI) or other survey administration methods [420].

There are two ways of connecting information generated by ICT-based data collection methods with planning activities: (1) by using ICT-based methods to collect or improve the data necessary for planning models, and (2) by using data generated by the pervasive sensing infrastructure directly for long-range planning and decision-making. GPS sensors have now been used to supplement several metropolitan-scale household data collection efforts. The major motivation for this trend has been to overcome data inaccuracies resulting from difficulties faced by respondents in comprehending survey diary questions, and in accurately recalling and self-reporting their travel patterns and activities. One particular accuracy concern is missing trips, when respondents forget to report trips that they have undertaken entirely. The model for data collection in such a scenario is to conduct a CATI survey of a larger sample of households and to request a subset of this larger sample to use GPS sensors to generate detailed information. In-vehicle or wearable GPS sensors have been used to passively collect detailed data on trips, or interactive methods, which allow the respondent to enter additional trip details via specialized computer interfaces, in addition to the passive GPS data collection [421] are used, among others.

Although the sensing systems within the primary tier of the DMII have the ability to generate large amounts of data, the processes and analytical tools by which these data streams can be seamlessly integrated into long-range transportation planning exist only on a limited scale at the level of practice, although at the research level, this issue is being actively considered. One active area of research is mobility mining which primarily utilizes massive amounts of GPS-based trajectory data to infer mobility patterns. This stream of work utilizes advancements in the areas of spatio-temporal data mining and geographic knowledge discovery to raw GPS data to uncover useful representations of mobility behaviors by means of mobility knowledge discovery process.

One example is the Geographic Privacy-aware Knowledge Discovery and Delivery (GeoKDD) project, the goal of which is to "develop theory, techniques and systems for geographic knowledge discovery, based on new privacy-preserving methods for extracting knowledge from large amounts of raw data referenced in space and time" [422]. It was noted that "there is a long way to go from raw data of individual trajectories up to high-level collective mobility knowledge, capable of supporting the decisions of mobility and transportation managers". Examples of geographic knowledge discovery from the GeoKDD project including discovery of popular itineraries, routes and departure times followed from the origin to the destination of people's travels and accessibility conditions to areas with large traffic attractors and other discoveries of interest are given in [423].

Models of the demand for travel explicitly consider human activities [424] (and interestingly defines travel as " …the medium by which an individual transports oneself between spatially dislocated activity participations"). They also consider traveler choice processes and have been stimulated partly by the need to estimate changes in travel behavior in response to innovative policies designed to achieve sustainability [425]. The highly active research area of travel behavior includes discrete choice modeling and microsimulation of human activity patterns. The activity-based modeling approach requires time-use survey data for analysis and the estimation and the collection of data regarding all in-home and out-of-home activities pursued by individuals over the course of a day or over multiple days.

Recent work involving GPS data include [426, 427]. The use of sensor data by means of these modeling approaches has the potential to yield new insights regarding ways in which travel outcomes and choice processes occur and will have particular value regarding ways in which the effects of ICT can be assessed and incorporated into planning practice. As such, they would be highly important for travel forecasting purposes. This line of work has the potential to be connected to planning processes at a future time.

4.4.3 ICT-Based Mobility Strategy Design for User-Centeredness and Sustainable Outcomes

Another dimension necessary to boost the ability of ICT to support sustainable travel is by considering the role of the user in the design of technologies. There is a need for explicit connection and feedback among different stages of a technology lifecycle and use-levels. The process of envisioning, designing, creating and using ICT form different stages of the ICT lifecycle. Yet research into specific aspects of the ICT lifecycle are concentrated in different academic disciplines and professional interest groups or industries, leaving little room for collaboration, cross-fertilization and feedback into future visioning and design stages [428]. This feedback loop should be researched so that broader outcome and impact evaluation results relating to ICT and travel can more explicitly inform specific product designs in the future.

Performance of a technology or systems of technologies based on evaluations at different use-levels can also serve to improve management practices and inform technology choices in the future. Use-levels can range from an individual device or person to an entire area, transportation corridor or a regional economy. Overall, the utility of ICT to derive economic, environmental and social sustainability outcomes has been assessed by numerous evaluations. This body of work to understand the "worth" of ICT-based mobility projects has used a variety of evaluation criteria, different units of analysis and has been conducted at different geographical levels.

Another aspect to user-centeredness and sustainable outcomes is the need to have a set of design principles that are based on a theoretical and empirical understanding of humans and their interactions and on design experiences and practices [428]. As noted by [429], design principles are high-level and largely context-free design goals which are fundamental, widely applicable and enduring, in contrast to design guidelines that give rules for designers to follow in order to design specific products. While broad-based research into this question is needed to derive principles that are inclusive and cover technologies for the vast range of functions that ICT-based mobility services currently cover—and will cover in the future—it is possible to imagine that a comprehensive set of design principles could include some or all of the following elements: safety and human factors; eco-design and eco-awareness; persuasion, support and user notification of sustainable and wellness alternatives to current activity; location privacy and information security; universal design for access by all; social connectedness and sharing; and user cost (time, dollar cost and so on due to use).

Chapter 5
Conclusions

5.1 Ubiquitous Information-Centered Mobility Environment

Transformations in wireless connectivity and location-aware technologies hold the promise of bringing a sea-change in the way transportation information will be generated and used in the future. Sensors in the transportation system, when integrated with those in other sectors (for example, energy, utility and health) have the potential to foster novel new ways of improving livability and sustainability. In this new transportation paradigm, the pervasive use of ICT will lead to a Digital Mobility Information Infrastructure (DMII) which will serve as the foundation for mobility intelligence towards an "ubiquitous information-centered mobility environment".

However, many technical and social challenges present themselves in the transformation to this vision. Interconnecting various domains will require computationally efficient methods to link and retrieve useful intelligence from unprecedented amounts of heterogeneous information ("Big Data"). Knowledge discovery and spatial data mining techniques will be necessary to extract information from massive datasets in real-time. Transportation engineering and planning models can utilize such information not only for improved system operations and management, but also to model and forecast sustainable development scenarios in the future.

In order to facilitate the equitable and efficient transformation of society to such an environment, there is a need to develop models of digital information governance, mobile digital citizenship and business innovation. There is also a need to study the various facets of the mobile digital divide and approaches to address this divide. Theoretical and computational models of dynamic pricing, insurance management and trading will be important to utilize information towards personal economic benefits, while incentives and user ratings will play an important role in motivating valid User-Generated Content. Human–Computer Interaction (HCI) principles for user-centered design of mobility technologies will be critical for increased and sustained use over time.

To introduce sustainable mobility and physical well-being considerations in opportunistic settings within daily travel, models of eco-feedback, persuasion and

P. (Vonu) Thakuriah and D. G. Geers, *Transportation and Information*, SpringerBriefs in Computer Science, DOI: 10.1007/978-1-4614-7129-5_5, © The Author(s) 2013

personal analytics can be useful in designing mobility technologies. Public acceptance may be determined partly by ways in which locational privacy, trust management and information security are considered and the extent to which resulting services are easy to use and provide cost-effective, reliable and safe mobility outcomes. Government agency or business acceptance will be determined partly by the extent to which ICT-based resource and asset management systems offer cost-efficiencies over existing models and on the availability of human resources, necessary tools and evaluation metrics.

Most of all, the question of whether this vision of a ubiquitous digital mobility environment is desirable or socially acceptable may still be still open. The DMII and resultant services should not only be the end-result of technological determinism, where the rationale is that desirable behaviors and outcomes will result simply by linking together existing or planned information assets, and by leveraging the very latest in ICT. Such a pragmatic and opportunistic rationale will undoubtedly be important to a certain extent in deriving novel and previously unknown ways of addressing mobility problems. However, for improved broad-based societal benefits, a strategic and deliberate approach, rather than one that is incremental and ad-hoc, is needed to organize and utilize the DMII. While a specific application area will have its own growth pattern, in order to strongly couple the links between ICT and sustainable mobility outcomes, the infrastructure should be a component of an overall regional mobility planning program that addresses investments in sustainable transportation alternatives (for example, shared and public transportation, walking, cycling, low-carbon personal vehicles and so on). At the same time, it should support information service development and experimentation towards knowledge-discovery and novel research questions.

5.2 Major Trends

Information pervasiveness in the DMII is the result of connected sensors, networks and their miniaturization; widespread proliferation of mobile technologies; reduction in the real costs of these connected systems; as well as information personalization and cooperation. The end-result of these developments have been somewhat contradictory. Although the level of automation in the transportation environment has greatly increased and autonomous environments have become widespread, the level of involvement and active participation by humans, in terms of co-creation and contribution of information, has also increased. The following two major trends can be observed regarding the DMII:

1. Increases in M2M communications;
2. Increases in the variety and volume of UGC.

As the volume of information in the mobile digital environment becomes increasingly large, these major trends will need to be supported by mobility analytics for:

1. Innovations in managing data and communications;

2. Novel applications to support transportation and mobility;
3. Diversity in tools and models to support information extraction, analysis and decision-making;
4. New models of governance, business, institutional and legal processes;
5. Analysis of previously unexplored social consequences of the digital mobility information environment;
6. Sustainable and user-centered technology design.

At the level of technology, two factors will be essential to support the ability of the DMII to foster sustainable outcomes. These are:

1. An *adequate infrastructure for mobility* (including sensor and communication infrastructure);
2. An *adequate information processing infrastructure* that includes data management and analysis capacity (software, tools, models) to extract and process relevant and timely mobility intelligence.

At the level of policy and practice, four factors will be essential for effective return-on-investment on the DMII:

1. *Coordination* between the DMII and plans for multi-modal accessible transportation, overall smart city initiatives and the use of ICT as a core element in shared and on-demand transportation;
2. *User-centeredness of technologies* that are sensitive to the attitudes and behavioral adaptations of users and which address the needs of a cross-section of the population to ensure their participation in the digital mobile environment;
3. *Governance, business practices and societal preparedness* including a legal infrastructure for privacy, trust and intellectual property management as well as strategies to expedite the level of digital citizenship and approaches to address issues of the mobile digital divide;
4. *Evaluation framework and operational metrics* for performance measurement regarding economic, environmental and social effectiveness which serves to continuously improve the design and operations of the DMII.

The above requirements have implications across a range of disciplines:

Computational Research: The fast-paced developments in sensor and social media technology opens up new sensing possibilities in transportation and mobility applications. Together with research into new directions in communications, positioning and information fusion, M2M communication at previously untested scales may be needed. Together, these trends open up the possibility for creating an information technology infrastructure for mobility-on-demand and shared transportation, leading to overall efficient use of resources. Privacy-by-design, trust management and information security will be key areas of research as vehicle-to-vehicle communication applications increase and the level of user contributions to the information infrastructure increases. Advances in sensor technology and augmented real world objects (augmented reality) are needed to facilitate assistive mobility technologies for persons with special needs (such as seniors or persons with disabilities) so they may seamlessly participate in everyday life as well; such research will also be needed for mHealth applications that support real-time decision-making.

Mobility Intelligence and Information Analysis: Data from Tier 1 of the DMII can be directly analyzed to produce mobility intelligence needed for transportation operations, management and service delivery. Information retrieval and extraction particularly for UGC and metrics for evaluating the quality of UGC are open areas of research within the transportation sphere and elsewhere. Scalability simulations for verifying the feasibility of communications in novel application scenarios will continue to be a needed. Data from the DMII can be imported into simulations and planning models for long-range transportation planning and scenario analysis; yet much research remains to be done in this area. The data can also be used to infer traffic and user behavior patterns for improved decision-support and to evaluate potential technology impacts.

Planning, Public Administration and Management Research: Research is needed on innovative models of information provision, novel business models, and models of governance, ownership and coordination related to managing information in the DMII. The overall trend towards a ubiquitous mobility information society can not only make transportation services of the future more efficient and reliable but can also transform existing business processes and stimulate entirely new business models. Transformative practices could emerge in B2B models in the LBS area, utility sectors or Smart City strategies, and in cross-cutting pricing applications, for example, dynamic pricing of location-aware services and real-time insurance policies.

Social Science Research: To understand the social consequences of the ubiquitous mobility information environment and implications for technology design, a discussion is needed to understand the limits, barriers, and saturation levels that may set in with "information overload" and to consider the needs of technology low-rate users or non-adopters and methods to ensure equitable access for all travellers.

Interdisciplinary Research on Human–Computer Interactions: A comprehensive set of user-centered design principles could facilitate technology adoption and sustainable use of transportation resources over time. These principles include those addressing safety and human factors; eco-design and eco-awareness; user notification of sustainable and wellness alternatives to the user's current travel choice; location privacy and information security; universal design for access by persons with disabilities; social connectedness and sharing; and considerations of user cost (time, dollar cost and so on). This stream is inherently interdisciplinary requiring cross-fertilization of ideas and research.

5.3 Conclusions

Transportation is a necessary and vital part of modern society. Without an efficient transportation network, economic growth and technological development would grind to a halt. However, the relentless drive to move ever more people and goods efficiently is hampered by the lack of urban space and concerns over cost and issues such as climate change. Society can no longer simply build additional infrastructure to meet increasing demand.

Hence alternative solutions have to be considered, as they already have, with the strong focus on ICT in transportation. ICT can play a range of supportive roles in transportation, for example, by supporting demand and asset management as well as assisting in operations. Many other supportive roles are possible. We have made the point in the book that travelers will increasingly play an important role in generating and sharing mobility information. This is a part of the overall trend towards developing "social" transportation systems where information and resource sharing with social networks (eg., on vehicles, itineraries) will be one component of sustainable mobility services. Social transportation systems of the future will need to be strongly coupled with dynamic resource management and mobility-on-demand strategies.

If ICT can improve the operational utility or extend the life of extant infrastructure by even a small amount then any investment in the technology required may be repaid many times over. However, the path forward is rocky. Technology moves ahead at a relentless pace and is lagged substantially by policy. New infrastructure should at the very least be ICT-ready if not enabled. Many of the outcomes of the DMII will occur at little or no cost as governments make more and more data freely available. V2V communications will be deployed by vehicle manufacturers as they strive to make their vehicles safer. The business case for wide-spread deployment of V2I is not clear. There is no doubt that road side units will be deployed at critical sites such as tunnel entrances, level-crossings and low load limit bridges but the deployment at every signalized junction in order to transmit signal and phase timing data to approaching vehicles, for example, seems much less likely.

Ultimately, the market will guide the types of applications and services that will be developed, whether to address the grey digital divide or to capitalize on innovations in location-aware technologies. Yet, there is much room for policy, as we have tried to stress in the book. The DMII and open data policies should continue to be encouraged for innovative applications, although the real benefits are yet to be evaluated systematically. Yet, desirable environmental, economic and socially sustainable outcomes may be possible to achieve when ICT in transportation is instituted within a broad-based strategy which includes, among other factors, coordination, governance and management innovation, appropriate pricing for system and service use, and societal preparedness. These are only some areas where the market can be guided to develop ICT mobility solutions that serve the interests of many.

References

1. International Road Federation. World road statistics 2002–2007. Technical report, 2009. Accessed on August 11, 2011.
2. Texas Transportation Institute. Urban mobility report 2010. Technical report, 2010. Accessed on August 11, 2011.
3. European Commission Mobility and Transport. Transport matters. http://ec.europa.eu/transport/strategies/facts-and-figures/transport-matters/index_en.htm, n.d.
4. UNEP and the FIA Foundation. Share the road: Investment in cycling and walking road infrastructure. Technical report, United Nations Environment Programme and FIA Foundation for the Automobile and Society, 2010. Accessed on August 29, 2012.
5. Energy Information Administration. International energy outlook 2010. Technical report, 2010. Access date: August 11, 2011, DOE/EIA-0484(2010).
6. International Traffic Safety Data and Analysis Group. Road safety 2011. oecd/itf 2011. Technical report, 2011. Access date: August 11, 2011.
7. Federal Commerce Commission. Internet access services: Status as of december 31, 2009. www.fcc.gov/Daily_Releases/Daily_Business/2010/db1208/DOC-303405A1.pdf, 2010. Accessed Jul. 30, 2011.
8. CTIA The Wireless Association. Wireless quick facts. www.ctia.org/advocacy/research/index.cfm/aid/10323, n.d. Accessed Jul. 30, 2011.
9. World Bank. The world bank data: Mobile cellular subscriptions (per 100 people). http://data.worldbank.org/, n.d.
10. D Shin. Ubiquitous city: Urban technologies, urban infrastructure and urban informatics. *Journal of Information Science*, 35(5):515–526, 2009.
11. Mark Weiser. The computer for the 21st century. *Scientific American*, 265(3):66–75, January 1991.
12. A Greenfield. *Everyware: The Dawning Age of Ubiquitous Computing*. New Riders Publishing, Berkeley, 2006.
13. J Innes and D E Booher. Public participation in planning: New strategies for the 21st century. Technical report, UC Berkeley, Institute of Urban and Regional Development Working Paper 2000–07, 2000. Accessed on August 8, 2011.
14. Douglas Schuler. *New community networks - wired for change*. Addison-Wesley, 1996.
15. M. Gurstein. *Community informatics: Enabling communities with information and communications technologies*. Idea Group Pub, 2000.
16. M. Foth, J.H. Choi, and C. Satchell. Urban informatics. In *Proc. ACM 2011 conference on Computer Supported Cooperative Work*, pages 1–8. ACM, 2011.
17. D Mont. Measuring disability prevalence. Technical Report Social Protection Discussion Paper No. 0706, March 2007. Access date: 08/01/2012.
18. World Bank. Sustainable transport: Priorities for reform. Technical report, 1996.
19. United Nation. Report of the world summit on sustainable development: Johannesburg, south africa. Technical report, 2002.

20. World Summit on the Information Society. Declaration of Principles: Building the Information Society: a global challenge in the new Millennium. http://www.itu.int/wsis/docs/geneva/official/dop.html, 2003. Access date: 08/20/2012.
21. International Telecommunication Union. Icts and climate change. Technical report, 2007.
22. R Weil, J Wootton, and A Garca-Ortiz. Traffic incident detection: Sensors and algorithms. *Mathematical and Computer Modelling*, 27(911):257–291, 1998.
23. J S Drake, J L Schofer, A May, A D May, Chicago Area Expressway Surveillance Project, and Expressway Surveillance Project (Ill.). *A Statistical Analysis of Speed-density Hypotheses: A Summary*. Report (Expressway Surveillance Project (Ill.)). Expressway Surveillance Project, 1965.
24. P J Lowrie. SCATS—Sydney Coordinated Adaptive Traffic System A Traffic Responsive Method of Controlling Urban Traffic. In *Proc. International Conference on Road Traffic Signaling, IEE, London, U.K.*, pages 67–70, 1982.
25. Hunt P B, Robertson D I, Bretherton R D, and Winton R I. SCOOT—a traffic responsive method of coordinating signals. Technical report, TRL Laboratory, Report 1014, 1981.
26. L A Klein, M K Mills, D R P Gibson, United States, and Turner-Fairbank Highway Research Center. *Traffic Detector Handbook*, volume 1. US Dept. of Transportation, Federal Highway Administration, Research, Development, and Technology, Turner-Fairbank Highway Research Center, McLean, VA, 3 edition, 2006.
27. M Su and J Luk. Evaluation of Sensys Networks Equipment: Stage 3 Tests on Monash Freeway. Technical report, ARRB, 2007.
28. S A Ahmed, T M Hussain, and T N Saadawi. Active and passive infrared sensors for vehicular traffic control. In *Vehicular Technology Conference, 1994 IEEE 44th*, volume 2, pages 1393–1397, jun 1994.
29. M Bao, C Zheng, X Li, J Yang, and J Tian. Acoustical Vehicle Detection Based on Bispectral Entropy. *Signal Processing Letters, IEEE*, 16(5):378–381, may 2009.
30. A Y Nooralahiyan, M Dougherty, D McKeown, and H R Kirby. A field trial of acoustic signature analysis for vehicle classification. *Transportation Research Part C: Emerging Technologies*, 5(34):165–177, 1997.
31. L A Klein, M K Mills, D R P Gibson, United States, and Turner-Fairbank Highway Research Center. *Traffic Detector Handbook*, volume 2. U.S. Dept. of Transportation, Federal Highway Administration, Research, Development, and Technology, Turner-Fairbank Highway Research Center, McLean, VA, 3 edition, 2006.
32. H Kim, J-H Lee, S-W Kim, J-I Ko, and D Cho. Ultrasonic vehicle detector for side-fire implementation and extensive results including harsh conditions. *Intelligent Transportation Systems, IEEE Transactions on*, 2(3):127–134, sep 2001.
33. M Skolnik. *Introduction to Radar Systems*. McGraw-Hill Science/Engineering/Math, 3 edition, 2002.
34. B Tian, Q Yao, Y Gu, K Wang, and Y Li. Video processing techniques for traffic flow monitoring: A survey. In *Intelligent Transportation Systems (ITSC), 2011 14th International IEEE Conference on*, pages 1103–1108, oct. 2011.
35. V Kastrinaki, M Zervakis, and K Kalaitzakis. A survey of video processing techniques for traffic applications. *Image and Vision Computing*, 21(4):359–381, 2003.
36. N Buch, S A Velastin, and J Orwell. A Review of Computer Vision Techniques for the Analysis of Urban Traffic. *Intelligent Transportation Systems, IEEE Transactions on*, 12(3):920–939, sept. 2011.
37. J Versavel. Road safety through video detection. In *Intelligent Transportation Systems, 1999. Proc. 1999 IEEE/IEEJ/JSAI International Conference on*, pages 753–757, 1999.
38. S M M Kahaki and M J Nordin. Vision-based automatic incident detection system using image sequences for intersections. In *Pattern Analysis and Intelligent Robotics (ICPAIR), 2011 International Conference on*, volume 1, pages 3–7, june 2011.

39. Y Zou, G Shi, H Shi, and Y Wang. Image Sequences Based Traffic Incident Detection for Signaled Intersections Using HMM. In *Hybrid Intelligent Systems, 2009. HIS '09. Ninth International Conference on*, volume 1, pages 257–261, aug. 2009.

40. G Kim, H Kim, J Park, J Kim, and Y Yu. Computationally Efficient Vehicle Tracking for Detecting Accidents in Tunnels. In T-H Kim, H Adeli, W I Grosky, N Pissinou, T K Shih, E J Rothwell, B-H Kang, and S-J Shin, editors, *Multimedia, Computer Graphics and Broadcasting, volume 263 of Communications in Computer and Information Science*, pages 197–202. Springer Berlin Heidelberg, 2011.

41. R C Gonzalez and P A Wintz. *Digital image processing*. Addison-Wesley Pub. Co., Advanced Book Program, Reading, Mass. : 1977.

42. M Piccardi. Background subtraction techniques: a review. In *Systems, Man and Cybernetics, 2004 IEEE International Conference on*, volume 4, pages 3099–3104, oct. 2004.

43. Y Benezeth, P M Jodoin, B Emile, H Laurent, and C Rosenberger. Review and evaluation of commonly-implemented background subtraction algorithms. In *Pattern Recognition, 2008. ICPR 2008. 19th International Conference on*, pages 1–4, dec. 2008.

44. K A Ahmad, Z Saad, N Abdullah, Z Hussain, and M H Mohd Noor. Moving vehicle segmentation in a dynamic background using self-adaptive kalman background method. In *Signal Processing and its Applications (CSPA), 2011 IEEE 7th International Colloquium on*, pages 439–442, march 2011.

45. H-T Zhai, J Wu, J Xia, and Z-M Cui. Self-adaptive Detection of Moving Vehicles in Traffic Video. In *Proc, 2009 International Symposium on Web Information Systems and Applications*, pages 449–453, May 2009.

46. A Prati, I Mikic, M M Trivedi, and R Cucchiara. Detecting moving shadows: algorithms and evaluation. *Pattern Analysis and Machine Intelligence, IEEE Transactions on*, 25(7): 918–923, july 2003.

47. A Prati, I Mikic, C Grana, and M M Trivedi. Shadow detection algorithms for traffic flow analysis: a comparative study. In *Intelligent Transportation Systems, 2001 Proc. 2001 IEEE*, pages 340–345, 2001.

48. Y Wang. Real-time moving vehicle detection with cast shadow removal in video based on conditional random field. *Circuits and Systems for Video Technology, IEEE Transactions on*, 19(3):437–441, march 2009.

49. K Robert. Night-Time Traffic Surveillance: A Robust Framework for Multi-vehicle Detection, Classification and Tracking. *Advanced Video and Signal Based Surveillance, IEEE Conference on*, 0:1–6, 2009.

50. R O Duda, P E Hart, and D G Stork. *Pattern Classification*. Wiley-Interscience, 2 edition, 2001.

51. V Vapnik. *The Nature of Statistical Learning Theory (Information Science and Statistics)*. Springer, 2 edition, 1999.

52. R E Schapire and Y Freund. *Boosting: Foundations and Algorithms (Adaptive Computation and Machine Learning series)*. The MIT Press, 2012.

53. R Hartley and A Zisserman. *Multiple View Geometry in Computer Vision*. Cambridge University Press, 2 edition, 2004.

54. Traficon N V. SafeWalk - The First Stereovision Video Sensor. http://www.traficon.com/pagenode.jsp?id=13&type=ProductCategory.

55. M C Schaefer. License Plate Matching Surveys: Practical Issues and Statistical Considerations. *ITE Journal*, 58(7):37–42, july 1988.

56. L A Klein and M R Kelleu. Detection Technology for IVHS Volume I: Final Report. http://ntl.bts.gov/lib/jpodocs/repts_te/6184.pdf, 1996.

57. I A Kostic, C E Holton, and R O Claus. Optical fiber magnetic field sensors and signal processing for vehicle detection and classification. volume 3525, pages 381–392. SPIE, 1999.

58. Y Wang, Q Wu, M Xiong, and R He. A new vehicle axle detector for roadways based on fiber optic Mach-Zehnder interferometer. volume 6019, page 60194F. SPIE, 2005.

59. W Li, F Zhu, and B Jiang. Technology of 0–20 ton FBG loadcell. volume 6781, page 67814P. SPIE, 2007.

60. K Wang, Z Wei, B Chen, and H-L Cui. A fiber-optic weigh-in-motion sensor using fiber Bragg gratings. volume 6004, pages 1–6. SPIE, 2005.

61. W J Fleming. Overview of automotive sensors. *Sensors Journal, IEEE*, 1(4):296–308, dec 2001.

62. R. Bishop. *Intelligent vehicle technology and trends*. Artech House, 2005.

63. R Bishop. Intelligent vehicle R &D: A review and contrast of programs worldwide and emerging trends. *Annals of Telecommunications*, 60:228–263, 2005. 10.1007/BF03219820.

64. N McKeegan. Volkswagen demonstrates fully-automatic reverse parking system. http://www.gizmag.com/volkswagen-demonstrates-fully-automatic-reverse-parking-system/9217/, 2008.

65. J Kwon and P Varaiya. Real-Time Estimation of Origin-Destination Matrices with Partial Trajectories from Electronic Toll Collection Tag Data. *Transportation Research Record: Journal of the Transportation Research Board*, 1923:119–126, 2005.

66. J S Wasson, J R Sturdevant, and D M Bullock. Real-Time Travel Time Estimates Using Media Access Control Address Matching. *ITE Journal*, 78(6):20–23, 2008.

67. University of Maryland-Center for Advanced Transportation Technology. Bluetooth Traffic Monitoring Technology. http://www.catt.umd.edu/documents/UMD-BT-Brochure_REV3.pdf, 2008.

68. N D Lane, E Miluzzo, Hong Lu, D Peebles, T Choudhury, and A T Campbell. A survey of mobile phone sensing. *Communications Magazine, IEEE*, 48(9):140–150, sept. 2010.

69. W Khan, Y Xiang, M Aalsalem, and Q Arshad. Mobile Phone Sensing Systems: A Survey. *Communications Surveys Tutorials, IEEE*, PP(99):1–26, 2012.

70. R K Rana, C T Chou, S S Kanhere, N Bulusu, and W Hu. Ear-phone: an end-to-end participatory urban noise mapping system. In *Proc. 9th ACM/IEEE International Conference on Information Processing in Sensor Networks*, IPSN '10, pages 105–116. ACM, 2010.

71. X Teng, Y Zhang, C C Y Poon, and P Bonato. Wearable medical systems for p-Health. *Biomedical Engineering, IEEE Reviews in*, 1:62–74, 2008.

72. J D Gibson, editor. *Mobile Communications Handbook*. CRC Press, 3 edition, 2012.

73. S Deering and R Hinden. *RFC 2460 Internet Protocol, Version 6 (IPv6) Specification*. Internet Engineering Task Force, December 1998.

74. International Organization for Standardization. Communications in Cooperative Intelligent Transport Systems—CALM for C-ITS. http://http://calm.its-standards.info/.

75. IEEE. IEEE 802.11™: Wireless LANs. http://standards.ieee.org/about/get/802/802.11.html, n.d.

76. H Hartenstein and K Laberteaux. *VANET: vehicular applications and inter-networking technologies*. Wiley Online, Library, 2010.

77. F. Dötzer. Privacy issues in vehicular ad hoc networks. In *Privacy Enhancing Technologies*, pages 197–209. Springer, 2006.

78. X. Lin, R. Lu, C. Zhang, H. Zhu, P.H. Ho, and X. Shen. Security in vehicular ad hoc networks. *Communications Magazine, IEEE*, 46(4):88–95, 2008.

79. VB Vaghela and DJ Shah. A survey and comparative study of vehicular traffic control system (vtcs). In *IJCA Proc. International Conference and workshop on Emerging Trends in Technology (ICWET 2012)*, number 12. Foundation of Computer Science (FCS), 2012.

80. European Commission. Co-operative systems for intelligent road safety. http://www.coopers-ip.eu/.

81. European Commission. Cooperative Vehicle Infrastructure Systems. http://www.cvisproject.org/.

82. European Commission. Preventive and Active Safety Applications. http://prevent.ertico.webhouse.net/.

83. European Commission. Cooperative Ssytems for Road Safety: Smart Vehicles on Smart Roads. http://www.safespot-eu.org/.

84. G Wu, S Talwar, K Johnsson, N Himayat, and K D Johnson. M2m: From mobile to embedded internet. *Communications Magazine, IEEE*, 49(4):36–43, april 2011.

85. Auto-id Labs. http://www.autoidlabs.org/.
86. ZigBee Alliance. http://www.zigbee.org/.
87. Borko Furht, editor. *Handbook of Augmented Reality*. Springer, 2011.
88. J W Hui and D E Culler. Extending IP to Low-Power, Wireless Personal Area Networks. *Internet Computing, IEEE*, 12(4):37–45, July-Aug. 2008.
89. N Gershenfeld and D Cohen. Internet 0: Interdevice Internetworking—End-to-End Modulation for Embedded Networks. *Circuits and Devices Magazine, IEEE*, 22(5):48–55, sept.-oct. 2006.
90. Living PlanIT. The Urban Operating System (UOS®). http://living-planit.com/UOS_overview.htm.
91. ZeroC. The Ice Manual. http://doc.zeroc.com/display/Ice/Ice+Manual.
92. R T Fielding. *Architectural styles and the design of network-based software architectures*. PhD thesis, 2000. AAI9980887.
93. RSS Advisory Board. RSS 2.0 Specification. http://www.rssboard.org/rss-specification.
94. RFC 4287 The Atom Syndication Format. Technical report, December 2005.
95. J Gregorio and B de hOra. The Atom Publishing Protocol, Internet RFC-5023, December 2007.
96. Transport for London. Developers' Area. http://www.tfl.gov.uk/businessandpartners/syndication/default.aspx.
97. Chicago Transport Authority. Developer Center. http://www.tfl.gov.uk/businessandpartners/syndication/default.aspx.
98. L E Sweeney Jr, W B Zavoli, and G Loughmiller. Comparative performance of various automotive navigation technologies. In *Vehicle Navigation and Information Systems Conference, 1993, Proc. IEEE-IEE*, pages 437–440, oct 1993.
99. G J Morgan-Owen and G T Johnston. Differential GPS positioning. *Electronics Communication, Engineering Journal*, 7(1):11–21, feb 1995.
100. P Enge, T Walter, S Pullen, Changdon Kee, Yi-Chung Chao, and Yeou-Jyh Tsai. Wide area augmentation of the Global Positioning System. *Proceedings of the IEEE*, 84(8): 1063–1088, aug 1996.
101. F Wu, N Kubo, and A Yasuda. Performance evaluation of GPS augmentation using Quasi-Zenith Satellite System. *Aerospace and Electronic Systems, IEEE Transactions on*, 40(4):1249–1260, oct. 2004.
102. A Varshavsky, Mike Y Chen, E de Lara, J Froehlich, D Haehnel, J Hightower, A LaMarca, F Potter, Timothy Sohn, K Tang, and I Smith. Are GSM Phones THE Solution for Localization? In *Mobile Computing Systems and Applications, 2006. WMCSA '06. Proc. 7th IEEE Workshop on*, pages 20–28, april 2006.
103. M Ibrahim and M Youssef. Cell Sense: An Accurate Energy-Efficient GSM Positioning System. *Vehicular Technology, IEEE Transactions on*, 61(1):286–296, jan. 2012.
104. S Feng and C L Law. Assisted GPS and its impact on navigation in intelligent transportation systems. In *Intelligent Transportation Systems, 2002. Proc. IEEE 5th International Conference on*, pages 926–931, 2002.
105. X Meng, L Yang, J Aponte, C Hill, T Moore, and A H Dodson. Development of Satellite Based Positioning and Navigation Facilities for Precise ITS Applications. In *Intelligent Transportation Systems, 2008. ITSC 2008. 11th International IEEE Conference on*, pages 962–967, oct. 2008.
106. R T Azuma. A Survey of Augmented Reality. *Presence*, 6:355–385, 1997.
107. J Guyot, G Falquet, and J Teller. Incremental development of a shared urban ontology: the urbamet experience. *Journal of Information Technology in Construction: Special Issue Bringing urban ontologies into practice*, pages 132–139, 2010.
108. C H Lau, Y Li, and D Tjondronegoro. Microblog retrieval using topical features and query expansion. http://trec.nist.gov/pubs/trec20/papers/QUT1.microblog.pdf, 2010.
109. D L Hall and J Llinas. An introduction to multisensor data fusion. *Proceedings of the IEEE*, 85(1):6–23, jan 1997.

110. C C Sun, G S Arr, R P Ramachandran, and S G Ritchie. Vehicle Reidentification using multidetector fusion. *Intelligent Transportation Systems, IEEE Transactions on*, 5(3): 155–164, sept 2004.

111. M Junghans and H-J Jentschel. Qualification of traffic data by Bayesian network data fusion. In *Information Fusion, 2007 10th International Conference on*, pages 1–7, july 2007.

112. P J Hargrave. A tutorial introduction to Kalman filtering. In *Kalman Filters: Introduction, Applications and Future Developments, IEE Colloquium on*, pages 1–6, feb 1989.

113. M S Arulampalam, S Maskell, and N Gordon. A tutorial on particle filters for online nonlinear/non-Gaussian Bayesian tracking. *IEEE Transactions on Signal Processing*, 50:174–188, 2002.

114. T Bass. Mythbusters: event stream processing versus complex event processing. In *DEBS '07: Proc. 2007 inaugural international conference on Distributed event-based systems*, volume 1, page 1, June 2007.

115. G Cugola and A Margara. Processing flows of information: From data stream to complex event processing. *ACM Comput. Surv.*, 44(3):1–62, June 2012.

116. G Ligozat, Z Vetulani, and J Osinski. Spatiotemporal Aspects of the Monitoring of Complex Events for Public Security Purposes. *Spatial Cognition and Computation*, 11(1):103–128, 2011.

117. T Abraham and J F Roddick. Survey of Spatio-Temporal Databases. *Geoinformatica*, 3:61–99, March 1999.

118. X Wang, X Zhou, and S Lu. Spatiotemporal data modelling and management: a survey. In *Technology of Object-Oriented Languages and Systems, 2000. TOOLS - Asia 2000. Proc. 36th International Conference on*, pages 202–211, 2000.

119. J F Roddick, E Hoel, M J Egenhofer, D Papadias, and B Salzberg. Spatial, temporal and spatio-temporal databases—hot issues and directions for phd research. *SIGMOD Rec.*, 33:126–131, June 2004.

120. N Pelekis, B Theodoulidis, I Kopanakis, and Y Theodoridis. Literature review of spatio-temporal database models. *The Knowledge Engineering Review*, 19(03):235–274, 2004.

121. A Carvalho, C Ribeiro, and A Augusto Sousa. A Spatio-temporal Database System Based on TimeDB and Oracle Spatial. In A Tjoa, L Xu, and S Chaudhry, editors, *Research and Practical Issues of Enterprise Information Systems, volume 205 of IFIP International Federation for Information Processing*, pp. 11–20. Springer, Boston, 2006. 10.1007/0-387-34456-X-2.

122. OSGeo Project. PostGIS. http://postgis.refractions.net/.

123. TerraLib. http://www.terralib.org/.

124. Open Geospatial Consortium. http://www.opengeospatial.org/.

125. A Guttman. R-trees: a dynamic index structure for spatial searching. *SIGMOD Rec.*, 14:47–57, June 1984.

126. J E Dobson and P F Fisher. Geoslavery. *Technology and Society Magazine, IEEE*, 22(1):47–52, 2003.

127. Tiancheng Li and Ninghui Li. On the tradeoff between privacy and utility in data publishing. In *Proc. 15th ACM SIGKDD International Conference on Knowledge Discovery and Data Mining*, pages 517–526, 2009.

128. Quantifying individuals trade-offs between privacy, liberty and security: The case of rail travel in uk. *Transportation Research Part A: Policy and Practice*, 44(3):169–181, 2010.

129. A R Beresford and F Stajano. Location privacy in pervasive computing. *Pervasive Computing, IEEE*, 2(1):46–55, Jan-Mar 2003.

130. M Gruteser and D Grunwald. Anonymous Usage of Location-Based Services Through Spatial and Temporal Cloaking. In *Proc. 1st international conference on Mobile systems, applications and services*, MobiSys '03, pages 31–42. ACM, 2003.

131. J Al-Muhtadi, R Campbell, A Kapadia, M D Mickunas, and S Yi. Routing through the mist: privacy preserving communication in ubiquitous computing environments. In *Proc. 22nd International Conference on Distributed, Computing Systems*, pp. 74–83, 2002.

132. David Goldschlag, Michael Reed, and Paul Syverson. Onion routing. *Commun. ACM*, 42(2): 39–41, Feb 1999.
133. N B Priyantha, A Chakraborty, and H Balakrishnan. The cricket location-support system. In *Proc. 6th International Conference on Mobile Computing and Networking*, MobiCom '00, pages 32–43, 2000.
134. D Da Silva, T Ann Kosa, S Marsh, and K El-Khatib. Examining privacy in vehicular ad-hoc networks. In *Proc. 2nd ACM International Symposium on Design and Analysis of Intelligent Vehicular Networks and Applications*, pages 105–110, 2012.
135. J Freudiger, M Raya, Felegyhazi, Papadimitratos M, P, and J-P Hubaux. Mix-zones for location privacy in vehicular networks. http://infoscience.epfl.ch/record/109437/files/FreudigerRFPH07winITS.pdf, 2007.
136. R Stewart and J Musters. Toronto: A historical leader in transportation innovations. *ITE Journal*, 68(4):30, 1998.
137. TRL Ltd. Transyt: Traffic network and isolated intersection study tool. https://www.trlsoftware.co.uk/products/junction_signal_design/transyt.
138. A Stevanovic. Adaptive Traffic Control Systems: Domestic and Foreign State of Practice. Technical Report SYN 403, Transportation Research Board, 2010.
139. NSW Roads and Maritime Services. Sydney Co-ordinated Adaptive Traffic System. http://www.scats.com.au/.
140. TRL Ltd. SCOOT-The world's leading adaptive traffic control system. http://www.scoot-utc.com/.
141. M McDonald. Road traffic control: TRANSYT and SCOOT. In *Concise Encyclopedia of Traffic & Transportation Systems (Advances in Systems Control and Information Engineering)*, pages 400–408. Pergamon Press, 1991.
142. C Kergaye, A Z Stevanovic, and P T Martin. An Evaluation of SCOOT and SCATS through Microsimulation. In *10th International Conference on Applications of Advanced Technologies in, Transportation*, May 2008.
143. A Engstrom. 10 years with LHOVRA—what are the experiences? In *Road Traffic Monitoring and Control, 1994, Seventh International Conference on*, pages 97–100, apr 1994.
144. P Mirchandani and L Head. A real-time traffic signal control system: architecture, algorithms, and analysis. *Transportation Research Part C: Emerging Technologies*, 9(6): 415–432, 2001.
145. V Mauro and D Di Taranto. "UTOPIA". In *Proc. 6th IFAC/IFIP/IFORS Symposium on Control and Communication in Transportation*, 1990.
146. B T Friedrich, M Sachse, W Hoops, and G Jendryschik. In *Proc. 2nd World Congressn on Intelligent Transportation Systems*, pages 2356–2361, 1995.
147. P Mirchandani and L Head. A real-time traffic signal control system: architecture, algorithms, and analysis. *Transportation Research Part C-emerging Technologies*, 9:415–432, 2001.
148. N H Gartner, F J Pooran, and Andrews C M. Optimized policies for adaptive control strategy in real-time traffic adaptive control systems: Implementation and field testing. *Transportation Research Record*, 1811:148–156, 2002.
149. F Crenshaw. ACS "Lite". http://www.signalsystems.org.vt.edu/documents/July00/Attach/Item03_Crenshaw_ACSLite.pdf, 2000.
150. R Ghaman, D Gettman, and S Shelby. ACS lite Project Overview. http://www.signalsystems.org.vt.edu/documents/Jan2004AnnualMeeting/SundayWorkshop/ACSLITE%20TRB%202004.pdf, 2004.
151. V Gradinescu, C Gorgorin, R Diaconescu, V Cristea, and L Iftode. Adaptive Traffic Lights Using Car-to-Car Communication. In *Vehicular Technology Conference, 2007 VTC2007-Spring. IEEE 65th*, pages 21–25, april 2007.
152. Investigating the Potential Benefits of Broadcasted Signal Phase and Timing (SPaT) Data under IntelliDrive ^{sm}. http://cts.virginia.edu/PFS_SPAT99_Final.pdf, 2011.

153. C Cai, Y Wang, and G Geers. Adaptive traffic signal control using wireless communications. In *Proc. 91st Annual Meeting of the Transportation Research Board*, January 2012. Paper 12–2998.

154. D P Masher, D W Ross, P J Wong, P L Tuan, H M Zeidler, and S Peracek. Guidelines for Design and Operating of Ramp Control Systems. Technical Report Standford Res. Inst. Rep. NCHRP 3–22, SRI Project 3340, SRI Menid Park, CA, 1975.

155. M Papageorgiou, H Hadj-Salem, and H M Blosseville. ALINEA: A local feedback control law for on-ramp metering. *Transportation Research Record*, 1320:58–64, 1991.

156. E B Kosmatopoulos and M Papageorgiou. Stability analysis of the freeway ramp metering control strategy ALINEA. 2003.

157. L Jacobsen, K Henry, and O Mahyar. Real-Time Metering Algorithm for Centralized Control. *Transportation Research Record: Journal of the Transportation Research Board*, 1232:17–26, 1989.

158. R Lau. Ramp Metering by Zone—The Minnesota Algorithm. Minnesota Department of Transportation, 1997.

159. G Paesani, J Kerr, P Perovich, and E Khosravi. System Wide Adaptive Ramp Metering In Southern California. In *Proc. 7th Annual Meeting, ITS America*, 1997.

160. L Corcoran and G Hickman. Freeway Ramp Metering Effects in Denver. In *59th Annual Meeting, Institute of Transportation Engineers*, 1989.

161. M Papageorgiou, J-M Blosseville, and H Haj-Salem. Modeling and real time con-trol of trac flow on the southern part of the Boulevard Peripherique in Paris—Part II: Coordinated on-ramp metering. *Transportation Research Part A*, 24:361–370, 1990.

162. I Papamichail, M Papageorgiou, V Vong, and J Gaffney. Heuristic Ramp-Metering Coordination Strategy Implemented at Monash Freeway, Australia. *Transportation Research Record*, 2178:10–20, 2010.

163. D Zabriezach and K Weegberg. Monash Freeway - at Warrigal Road - Peak Period Traffic and Crash Data. Private, Communication, 2012.

164. European Commission. Eropean Rsmp Metering Project. http://www2.napier.ac.uk/euramp/home.htm, 2007.

165. M-C Esposito, J W Polak, R Krishnan, and M Pleydell. A global comparison of ramp-metering algorithms optimising traffic distribution on motorways and arterials. In *Road Transport Information and Control Conference and the ITS United Kingdom Members' Conference (RTIC 2010) - Better transport through technology, IET*, pages 1–6, may 2010.

166. X-J Xing-Ju Wang, X-M Xiao-Ming Xi, and G-F Gao. Reinforcement Learning Ramp Metering without Complete Information. *Journal of Control Science and Engineering*, 2012.

167. Z Hou, J-X Xu, and H Zhong. Freeway Traffic Control Using Iterative Learning Control-Based Ramp Metering and Speed Signaling. *Vehicular Technology, IEEE Transactions on*, 56(2):466–477, march 2007.

168. X-Y Lu, T Z Qiu, P Varaiya, R Horowitz, and S E Shladover. Combining Variable Speed Limits with Ramp Metering for freeway traffic control. In *American Control Conference (ACC)*, 2010, pages 2266–2271, 30 2010-july 2 2010.

169. X-Y Lu, P Varaiya, R Horowitz, Dongyan Su, and S E Shladover. A new approach for combined freeway Variable Speed Limits and Coordinated Ramp Metering. In *Intelligent Transportation Systems (ITSC), 2010 13th International IEEE Conference on*, pages 491–498, sept 2010.

170. A Hegyi, B De Schutter, and H Hellendoorn. Model predictive control for optimal coordination of ramp metering and variable speed limits. *Transportation Research Part C*, 13(3):185–209, June 2005.

171. L D LBaskar, B De Schutter, and H Hellendoorn. Dynamic speed limits and on-ramp metering for IVHS using model predictive control. In *Proc. 11th International IEEE Conference on Intelligent Transportation Systems (ITSC 2008)*, pages 821–826, October 2008.

172. Z Tian. Modeling and Implementation of an Integrated Ramp Metering-Diamond Interchange Control System. *J Transpn Sys Eng & IT*, 7(1):61–72, 2007.
173. C A MacCarley, S P Mattingly, M G McNally, D Mezger, and J E Moore. Field Operational Test of Integrated Freeway Ramp Metering/Arterial Adaptive Signal Control: Lessons Learned in Irvine, California. *Transportation Research Record: Journal of the Transportation Research Board*, 1811:79–83, 2002.
174. Z Z Tian, K Balke, R Engelbrecht, and L Rilett. Integrated Control Strategies for Surface Street and Freeway Systems. *Transportation Research Record: Journal of the Transportation Research Board*, 1811:92–99, 2002.
175. M van den Berg, A Hegyi, B De Schutter, and J Hellendoorn. Integrated traffic control for mixed urban and freeway networks: A model predictive control approach. *European Journal of Transport and Infrastructure Research*, 7(3):223–250, September 2007.
176. Federal Highway Administration. Traffic Incident Management. http://ops.fhwa.dot.gov/aboutus/one_pagers/tim.htm.
177. D Judyicki and J R Robinson. Freeway incident management. Technical report, Office of Traffic Operations, Federal Highway Administration, 1988.
178. B N Persaud and F L Hall. Catastrophe theory and patterns in 30-second freeway traffic data implications for incident detection. *Transportation Research Part A: General*, 23(2): 103–113, 1989.
179. K Ozbay and P Kachroo. *Incident Management in Intelligent Transportation Systems*. Artech House, 1999.
180. K. Ozbay and P. Kachroo. Incident management in intelligent transportation systems. 1999.
181. M Mancini, S Mitrovich, and G Valenti. A decision support system (dss) for traffic incident management in roadway tunnel infrastructure. In *European Transport Conference 2006*, 2006.
182. E H C Choi, R Taib, Y Shi, and F Chen. Multimodal user interface for traffic incident management in control room. *Intelligent Transport Systems, IET*, 1(1):27–36, march 2007.
183. Park Assist. Park Assist Launches the Next Generation Parking Guidance System. http://www.parkassist.com/index.php/about/news/adding-a-new-post/.
184. H-L Zhong, J-M Xu, Y Tu, Y-C Hu, and J-F Sun. The research of parking guidance and information system based on dedicated short range communication. In *Intelligent Transportation Systems, 2003. Proc. 2003 IEEE*, volume 2, pages 1183–1186, oct. 2003.
185. Z Yang, H Liu, and X Wang. The research on the key technologies for improving efficiency of parking guidance system. In *Intelligent Transportation Systems, 2003. Proc. 2003 IEEE*, volume 2, pages 1177–1182, oct. 2003.
186. J Yang. A System Framework of Active Parking Guidance and Information System. In *Information Engineering (ICIE), 2010 WASE International Conference on*, volume 2, pages 150–154, aug. 2010.
187. Star Tribune. 13 Seconds in August. http://www.startribune.com/local/12166286.html, 2007.
188. H. Sohn, C R Farrar, F M Hemez, D D Shunk, D W Stinemates, B R Nadler, and J J Czarnecki. *A review of structural health monitoring literature: 1996–2001*. Los Alamos National Laboratory Los Alamos, New Mexico, 2004.
189. C Yu, J Wang, L Tan, and X Tu. A bridge structural health data analysis model based on semi-supervised learning. In *Automation and Logistics (ICAL), 2011 IEEE International Conference on*, pages 30–34, aug. 2011.
190. P C Chang, A Flatau, and S C Liu. Review Paper: Health Monitoring of Civil Infrastructure. In *Structural Health Monitoring 2003: From Diagnostics & Prognostics to Structural Health Management: Proc. 4th International Workshop on, Structural Health Monitoring*, 2003.
191. D. INAUDI. Overview of 40 bridge structural health monitoring projects. *SMARTEC SA, Switzerland, IBC*, pages 09–45, 2010.
192. Y Li and W Zhao. Intelligent bridge monitoring system based on 3G. In *Consumer Electronics, Communications and Networks (CECNet), 2011 International Conference on*, pages 426–429, april 2011.

193. T Harms, S Sedigh, and F Bastianini. Structural Health Monitoring of Bridges Using Wireless Sensor Networks. *Instrumentation Measurement Magazine, IEEE*, 13(6):14–18, december 2010.

194. K D Janoyan and M J Whelan. Intelligent Transportation Infrastructure technologies for condition assessment and structural health monitoring of highway bridges. In K Gopalakrishnan and S Peeta, editors, *Sustainable and Resilient Critical Infrastructure Systems*, pages 159–184. Springer Berlin Heidelberg, 2010.

195. J A Rice and B L Spencer, Jr. Flexible Smart Sensor Framework for Autonomous Full-scale Structural Health Monitoring. Technical Report NSEL-018, University of Illinois at Urbana-Champaign, 2009.

196. NICTA. TruckOn. http://www.nicta.com.au/business/itl/projects/truckon.

197. N Brohlin. A Statistical Analysis of 28,000 Accident Cases with Emphasis on Occupant Restraint Value. Technical Report Technical Paper 670925, SAE International, 1967.

198. M Bellis. The History of Airbags. http://inventors.about.com/od/astartinventions/a/air_bags.htm, n.d.

199. Toyota Develops Advanced Head Restraint to Reduce Whiplash Injuries. http://www.autospies.com/news/Toyota-Develops-Advanced-Head-Restraint-to-Reduce-Whiplash-Injuries-17313/, 2007.

200. D Gazis, R Herman, and A Maradudin. The Problem of the Amber Signal Light in Traffic Flow. *Operations Research*, 8(1):112–132, 1960.

201. The Vision Zero Initiative. http://www.visionzeroinitiative.com/en/, n.d.

202. M Kakauskas, J Creaser, M Manser, J Graving, and M Donath. Validation Study—On-Road Evaluation of the Stop Sign Assist Decision Support Sign. Technical Report CICAS-SSA Report #5, University of Minnesota, 2009.

203. H A B S Works. Antilock braking systems.

204. L Austin and D Morrey. Recent advances in antilock braking systems and traction control systems. *Proc. Inst. of Mechanical Engineers, Part D: Journal of Automobile, Engineering*, 214(6):625–638, 2000.

205. C M Rudin-Brown and P C Burns. The secret of electronic stability control (esc). In *Canadian Multidisciplinary Road Safety Conference, 17th, 2007, Montreal, Quebec, Canada*, 2007.

206. G Bahouth. Real world crash evaluation of vehicle stability control (vsc) technology. In *Annual Proceedings/Association for the Advancement of Automotive Medicine*, volume 49, page 19. Association for the Advancement of Automotive Medicine, 2005.

207. G Buschmann, H T Ebner, and W Kuhn. Electronic brake force distribution control-a sophisticated addition to abs. *SAE paper*, 920646, 1992.

208. K Koibuchi, M Yamamoto, Y Fukada, and S Inagaki. Vehicle stability control in limit cornering by active brake. *SAE paper*, 960487, 1996.

209. J Gaus. The future of vehicle safety. Automobile Engineering International, pages 182–183, April 2005.

210. Cohda Wireless Pty. Ltd. Personal Communication.

211. Us, DOT. Safety Pilot. http://www.its.dot.gov/safety_pilot/.

212. A Desai, J Singh, and T Spicer. Intelligent Transport System to Improve Safety at Road-Rail Crossings. In *11th World Level Crossing Symposium*.

213. European Commission. eCall: Time saved = lives saved. http://ec.europa.eu/information_society/activities/esafety/ecall/index_en.htm.

214. S G Klauer, T A Dingus, V L Neale, J D Sudweeks, and D J Ramsey. The impact of driver inattention on near-crash/crash risk: An analysis using the 100-car naturalistic driving study data. Technical Report DOT HS 810 594, National Highway Traffic Safety Administration, 2006.

215. World Health Organization. Global status report on road safety: Time for action. Technical report, 2009. Access date: 08/12/2012.

216. R N Hardy, J W Watson, and K Kayvantash. Safety requirements for cyclists during leg impacts. In *Proc. 21st ESV International Technical Conference*, 2009.
217. M de Jong. SAFECYCLE: enhancing safety of cyclists with ICT. http://www.polisnetwork.eu/uploads/Modules/PublicDocuments/safecycleshort02.pdf, November 2011. (Access date: 08/12/2012).
218. D Gavrila. Pedestrian detection from a moving vehicle. In D Vernon, editor, *Computer Vision: ECCV 2000*, volume 1843 of *Lecture Notes in Computer Science*, pages 37–49. Springer Berlin/Heidelberg, 2000.
219. A Shashua, Y Gdalyahu, and G Hayun. Pedestrian detection for driving assistance systems: single-frame classification and system level performance. In *Intelligent Vehicles Symposium, 2004 IEEE*, pages 1–6, June 2004.
220. D Gerónimo, A M López, A D Sappa, and T Graf. Survey of pedestrian detection for Advanced Driver Assistance Systems. *Pattern Analysis and Machine Intelligence, IEEE Transactions on*, 32(7):1239–1258, July 2010.
221. H Cho, P E Rybski, and W Zhang. Vision-based bicycle detection and tracking using a deformable part model and an EKF algorithm. In *Intelligent Transportation Systems (ITSC), 2010 13th International IEEE Conference on*, pages 1875–1880, Sept 2010.
222. M Bielli, A Bielli, and R Rossi. Trends in models and algorithms for fleet management. *Procedia - Social and Behavioral Sciences*, 20(0):4–18, 2011.
223. E Taniguchi and R G Thompson. Modeling City Logistics. *Transportation Research Record: Journal of the Transportation Research Board*, 1790:45–51, 2002.
224. A Glaschenko, A Ivaschenko, G Rzevski, and P Skobelev. Multi-Agent real time scheduling system for taxi companies. In *AAMAS 2009 - 8th International Conference on Autonomous Agents and Multiagent Systems*, 2009.
225. P Vogel and D Mattfeld. Strategic and operational planning of bike-sharing systems by data mining: A case study. In J W Böe, H Hu, C Jahn, X Shi, R Stahlbock, and S Voß, editors, *Computational Logistics*, volume 6971 of *Lecture Notes in Computer Science*, pages 127–141. Springer Berlin/Heidelberg, 2011.
226. P Thakuriah and N Tilahun. Incorporating Weather Information into Real-Time Speed Estimates: Comparison of Alternative Models. *Journal of Transportation Engineering*, 139(4):379–389, 2013.
227. S Drobot, W P Mahoney III, E Schuler, G Wiener, M Chapman, P A Pisano, P Kennedy, B B McKeever, and A D Stern. IntelliDrive SM road weather research and development - the vehicle data translator. Technical report, Federal Highway Administration Road Weather Management Program, 2009. Access date: 08/01/2012.
228. M Demirbas, M A Bayir, C G Akcora, Y S Yilmaz, and H Ferhatosmanoglu. Crowd-sourced sensing and collaboration using twitter. In *World of Wireless Mobile and Multimedia Networks (WoWMoM), 2010 IEEE International Symposium on a*, pages 1–9. IEEE, 2010.
229. H S Mahmassani, Dong J, Kim J, Chen R B, and Park B. Incorporating weather impacts in traffic estimation and prediction systems. Technical Report FHWA-JPO-09-065, Research and Technology Administration, ITS Joint Program, Office, 2009.
230. C Wattegama. ICT for Disaster Management. Technical report, UNDP Asia-Pacific Development Information Programme (UNDP-APDIP), 2007. Access date: 8/01/2012.
231. A Berube, E Deakin, and S Raphael. Socioeconomic differences in household automobile ownership rates: Implications for evacuation policy. In J M Quigley and L A Rosenthal, editors, *Risking house and home: disasters, cities, public policy*, pages 197–222. Berkeley Public Policy Press, Institute of Governmental Studies Publications, 2008. Access date: 08/10/2012.
232. B R Lindsay. Social media and disasters: Current uses, future options, and policy considerations. Technical Report 7-5700, Congressional Research Service, 2011. Access date: 08/01/2012.

233. I Shklovski, L Palen, and J Sutton. Finding community through information and communication technology in disaster response. In *Proc. ACM Conference on Computer Supported Cooperative Work (CSCW)*, pages 127–136, 2008.

234. C H. Procopio and S T Procopio. Do you know what it means to miss New Orleans? internet communication, geographic community, and social capital in crisis. *Journal of Applied Communication Research*, 35(1):67–87, 2007.

235. C Hagar. Using research to aid the design of a crisis information management course. Paper presented at the ALISE SIG Multicultural, Ethnic and Humanistic Concerns (MEH) session on Information Seeking and Service Delivery for Communities in Disaster/Crisis, 2006.

236. L Palen, S Vieweg, S B Liu, and A L Hughes. Crisis in a networked world. *Social Science Computer Review*, 27(4):467–480, Nov 2009.

237. K M Anderson and A Schram. Design and implementation of a data analytics infrastructure in support of crisis informatics research (nier track). In *Proc. 33rd International Conference on Software Engineering (ICSE)*, pages 844–847, 2011.

238. A Christensen. In the shadow of Copenhagen: Combining wind power and electric vehicles in Denmark to address climate change. http://asmarterplanet.com/blog/2009/12/wind-power-and-electric-vehicles-in-denmark.html, Dec. 2009. Access date: 08/10/2012.

239. M Naphade, G Banavar, C Harrison, J Paraszczak, and R Morris. Smarter Cities and their innovation challenges. *Computer*, 44(6):32–39, June 2011.

240. W Kempton and S E Letendre. Electric vehicles as a new power source for electric utilities. *Transportation Research Part D: Transport and Environment*, 2(3):157–175, 1997.

241. W Kempton and J Tomić. Vehicle-To-Grid power fundamentals: Calculating capacity and net revenue. *Journal of Power Sources*, 144(1):268–279, 2005.

242. C Bottrill. Understanding DTQs and PCAs. Technical report, Environmental Change Institute/UKERC, October 2006.

243. Doug Lundquist. Pollution credit trading in Vehicular Ad Hoc Networks. http://connected vehicle.challenge.gov/submissions/2926-pollution-credit-trading-in-vehicular-ad-hoc-networks, 2011.

244. MIT Media Lab. Mobility on Demand: Future of transportation in cities. Technical report, Smart Cities MIT Media Laboratory, 2008.

245. J Lee, S Baik, and C Lee. Building an integrated service management platform for ubiquitous cities. *Computer*, 44(6):56–63, June 2011.

246. B Liu, D Ghosal, Y Dong, C Chuah, and M Zhang. CarbonRecorder: A mobile-social vehicular carbon emission tracking application suite. In *Vehicular Technology Conference (VTC Fall), 2011 IEEE*, pages 1–2, Sept 2011.

247. MIT SENSEable City Lab. CO2GO. http://senseable.mit.edu/co2go/, n.d. Access date: 08/10/2012.

248. C Endres, J Miksatko, and D Braun. Youldeco - exploiting the power of online social networks for eco-friendly driving. In *Adjunct Proc. 2nd International Conference on Automotive User Interfaces and Interactive Vehicular Applications (AutomotiveUI 2010)*, volume 5, November 2010.

249. M Barth and K Boriboonsomsin. Energy and emissions impacts of a freeway-based dynamic eco-driving system. *Transportation Research Part D: Transport and Environment*, 14(6):400–410, 2009.

250. Read-Ahead Packet AERIS User Needs Workshop March 14–15th. AERIS transformative concepts: "Cleaner Air Through Smarter Transportation" eco-signal operations. Technical report, Research and Technology Administration, ITS Joint Program, Office, 2012.

251. D Toppeta. The Smart City vision: How innovation and ict can build smart, livable, sustainable cities. Technical report, The Innovation Knowledge Foundation, 2010. Access date: 08/01/2012.

252. A Caragliu, C Del Bo, and P Nijkamp. Smart Cities in europe. *Journal of Urban Technology*, 18(2):65–82, 2011.

253. R Bonney, H Ballard, R Jordan, E McCallie, T Phillips, J Shirk, and C C Wilderman. Public participation in scientific research: Defining the field and assessing its potential for informal science education. Technical report, Center for Advancement of Informal Science Education, 2009. Access date: 08/01/2012.

254. P S Kidd and M B Parshall. Getting the Focus and the Group: Enhancing analytical rigor in focus group research. *Qualitative Health Research*, 10(3):293–308, 2000.

255. R Jungk and N Müller. Future Workshops: How to create desirable futures. Technical report, 1987.

256. R V V Vidal. Scenario Methods and Applications. Technical report, 1996.

257. C Eden. Strategic options development and analysis - soda. pages 21–42, 1989.

258. P Checkland. Systems Thinking, *Systems Practice*. John Wiley & Sons, 1981.

259. S D N Graham. Flight to the cyber suburbs. pages 2–3, April 1996.

260. L von Ahn, M Blum, N J. Hopper, and J Langford. CAPTCHA: Using hard AI problems for Security. Technical, Report 136, 2003.

261. L von Ahn, B Maurer, C McMillen, D Abraham, and M Blum. reCAPTCHA: Human-based character recognition via web security measures. *Science*, 321(5895):1465–1468, 2008.

262. M F. Goodchild. *Citizens as Sensors: The World of Volunteered Geography*, pages 370–378. John Wiley & Sons, Ltd., 2011.

263. K P Tang, J Lin, J I Hong, D P. Siewiorek, and N Sadeh. Rethinking location sharing: exploring the implications of social-driven vs. purpose-driven location sharing. In *Proc. 12th ACM International Conference on Ubiquitous computing (Ubicomp '10)*, pages 85–94, 2010.

264. A J Quinn and B B Bederson. Human computation: a survey and taxonomy of a growing field. In *Proc. Annual Conference on Human factors in computing systems (CHI '11)*, pages 1403–1412, 2011.

265. B Liu. *Web Data Mining: Data-Centric Systems and Applications*. Springer, 2007.

266. K Sasaki, S Nagano, K Ueno, and K Cho. Feasibility study on detection of transportation information exploiting Twitter as a sensor. 2012.

267. Twitter. Rest api v1.1 resources. https://dev.twitter.com/docs/api/1.1, n.d.

268. D Pfoser. On User-Generated Geocontent. In Dieter Pfoser, Yufei Tao, Kyriakos Mouratidis, Mario Nascimento, Mohamed Mokbel, Shashi Shekhar, and Yan Huang, editors, *Advances in Spatial and Temporal Databases*, volume 6849 of *Lecture Notes in Computer Science*, pages 458–461. Springer, 2011. 10.1007/978-3-642-22922-0-30.

269. K S Hornsby and N Li. Conceptual Framework for Modeling Dynamic Paths from Natural Language Expressions. *Transactions in GIS*, 13:27–45, 2009.

270. B Han. Reasoning about a temporal scenario in natural language. In *Proc. IJCAI Workshop on Spatial and Temporal Reasoning*, pages 19–26, july 2009.

271. N D Lane, S B Eisenman, M Musolesi, E Miluzzo, and A T Campbell. Urban sensing systems: opportunistic or participatory? In *Proc. 9th workshop on Mobile computing systems and applications*, HotMobile '08, pages 11–16, 2008.

272. A Namatame, S Kurihara, and H Nakashima, editors. *Emergent Intelligence of Networked Agents*, volume 56 of *Studies in Computational Intelligence*. Springer, 2007.

273. Y Ren, R Kraut, and S Kiesler. Applying common identity and bond theory to design of online communities. *Organization Studies*, 28(3):377–408, 2007.

274. E J. Friedman and P Resnick. The social cost of cheap pseudonyms. *Journal of Economics & Management Strategy*, 10(2):173–199, 2001.

275. Patricia Wallace. *The Psychology of the Internet*. Cambridge University Press, 1999.

276. R Jenkins. Categorization: Identity, social process and epistemology. *Current Sociology*, 48(3):7–25, 2000.

277. J Code and N Zaparyniuk. Social identities, group formation, and analysis of online communities. In *Dasgupta, S (Ed.), Social Computing: Concepts, Methodologies, Tools, and Applications*, pages 1346–1361. IGI Publishing, New York, NY, 2010.

278. I Varlamis, M Eirinaki, and M Louta. A study on social network metrics and their application in trust networks. In *Advances in Social Networks Analysis and Mining (ASONAM), 2010 International Conference on*, pages 168–175, Aug 2010.

279. S Ma, O Wolfson, and J Lin. A survey on trust management for intelligent transportation system. In *Proc. 4th ACM SIGSPATIAL International Workshop on Computational Transportation Science*, CTS '11, pages 18–23. ACM, 2011.

280. Sasank Reddy, Deborah Estrin, Mark Hansen, and Mani Srivastava. Examining micropayments for participatory sensing data collections. In *Proc. 12th ACM international Conference on Ubiquitous computing*, Ubicomp '10, pages 33–36. ACM, 2010.

281. P Thakuriah, A Sen, and A Karr. Probe-based surveillance for travel time information in ITS. In R Emmerink and P Nijkamp, editors, *Behavioral and Network Impacts of Driver Information Systems*, pages 393–425. Ashgate Publishing Ltd., 1999.

282. F Dion, R Robinson, and J Oh. Evaluation of usability of intellidrive probe vehicle data for transportation systems performance analysis. *Journal of Transportation Engineering*, 137(3):174–183, 2011.

283. E Agichtein, C Castillo, D Donato, A Gionis, and G Mishne. Finding high-quality content in social media. In *Proc. International Conference on Web search and web data mining*, WSDM '08, pages 183–194, 2008.

284. N Weber and S N Lindstaedt. A user centered approach for quality assessment in social systems. In *KMIS'11*, pages 211–216, 2011.

285. J Bleiholder and F Naumann. Data fusion. *ACM Comput. Surv.*, 41(1):1:1–1:41, January 2009.

286. M Anwar Hossain, P K Atrey, and A E Saddik. Modeling and assessing quality of information in multisensor multimedia monitoring systems. *ACM Trans. Multimedia Comput. Commun. Appl.*, 7(1):3:1–3:30, 2011.

287. D Hecker, H Stange, C Korner, and M May. Sample bias due to missing data in mobility surveys. In *Proc. IEEE International Conference on Data Mining Workshops (ICDMW '10)*, pages 241–248, 2010.

288. A Ghose and S P Han. An empirical analysis of user content generation and usage behavior on the mobile internet. *Manage. Sci.*, 57(9):1671–1691, 2011.

289. L Daz, C Granell, and J Huerta. IGI Global, 2012.

290. J Ding, L Gravano, and N Shivakumar. Computing geographical scopes of web resources. In *Proc. 26th International Conference on Very Large Data Bases (VLDB '00)*, pages 545–556. Morgan Kaufmann Publishers Inc., 2000.

291. A Markowetz, T Brinkhoff, and B Seeger. Geographic information retrieval. Access date: 08/01/2012.

292. C Sallaberry, M Gaio, J Lesbegueries, and P Loustau. A semantic approach for geospatial information extraction from unstructured documents. In A Scharl and K Tochtermann, editors, *The Geospatial Web*, Advanced Information and Knowledge Processing, pages 93–104. Springer London, 2009.

293. D Ahlers and S Boll. Urban web crawling. In *Proc. 1st International Workshop on Location and the Web (LOCWEB '08)*, pages 25–32, 2008.

294. S Timpf. Ontologies of wayfinding: a traveler's perspective. *Networks and spatial economics*, 2(1):9–33, 2002.

295. R Llyod. *Spatial Cognition: Geographic Environments*. The GeoJournal Library. Kluwer Academic Publishers, 1997.

296. S Pallottino and M G Scutella. Shortest path algorithms in transportation models: classical and innovative aspects. Technical report, 1997.

297. I C M Flinsenberg. *Route Planning Algorithms for Car Navigation*. PhD thesis, Technische Universiteit Eindhoven, 2004.

298. E I Vlahogianni, J C Golias, and M G Karlaftis. Short-term traffic forecasting: Overview of objectives and methods. *Transport Reviews*, 24(5):533–557, 2004.

299. S Peeta and Ziliaskopolous. *Foundations of Dynamic Traffic Assignment: The Past, the Present and the Future*, pages 233–265. Kluwer Academic Publishers, 2001.

300. W Y Szeto and H K Lo. Dynamic traffic assigment: Properties and extensions. *Transportmetrica*, 2(1):31–52, 2006.

301. M Brereton, P Roe, M Foth, J M Bunker, and L Buys. Designing participation in agile ridesharing with mobile social software. In *Proc. 21st Annual Conference of the Australian Computer-Human Interaction Special Interest Group: Design: Open 24/7*, OZCHI '09, pages 257–260, 2009.

302. R Trasarti, F Pinelli, M Nanni, and F Giannotti. Mining mobility user profiles for car pooling. In *Proc. 15th ACM SIGKDD International Conference on Knowledge Discovery and Data Mining (KDD '11)*, pages 1190–1198, 2011.

303. J Auld and A Mohammadian. Activity planning processes in the agent-based dynamic activity planning and travel scheduling (adapts) model. *Transportation Research Part A: Policy and Practice*, 46(8):1386–1403, 2012.

304. Y Zheng. Location-Based Social Networks: Users. In Y Zheng and X Zhou, editors, *Computing with Spatial Trajectories*, pages 243–276. Springer New York, 2011.

305. D Jannach, M Zanker, A Felfernig, and G Friedrich. *Recommender Systems: An Introduction*. Cambridge University Press, 2010.

306. A Tveit. Peer-to-peer based recommendations for mobile commerce. In *Proc. 1st International Workshop on Mobile commerce (WMC '01)*, pages 26–29, 2001.

307. Y Ge, H Xiong, A Tuzhilin, K Xiao, M Gruteser, and M Pazzani. An energy-efficient mobile recommender system. In *Proc. 16th ACM SIGKDD International Conference on Knowledge Discovery and Data Mining (KDD '10)*, pages 899–908, 2010.

308. B Ludwig, B Zenker, and J Schrader. Recommendation of personalized routes with public transport connections. In D Tavangarian, T Kirste, D Timmermann, U Lucke, and D Versick, editors, *Intelligent Interactive Assistance and Mobile Multimedia Computing*, volume 53 of *Communications in Computer and Information Science*, pages 97–107. Springer, 2009.

309. Y Zheng and X Xie. Location-Based Social Networks: Locations. In Y Zheng and X Zhou, editors, *Computing with Spatial Trajectories*, pages 277–308. Springer New York, 2011.

310. S Consolvo, K Everitt, I Smith, and J A Landay. Design requirements for technologies that encourage physical activity. In *Proc. SIGCHI Conference on Human Factors in computing systems (CHI '06)*, pages 457–466, 2006.

311. Nokia Research Center (NRC). Sensing the world with mobile devices: The vision. Technical report, December 2008. Access date: 08/01/2012.

312. S B Eisenman, E Miluzzo, N D Lane, R A Peterson, G Ahn, and A T Campbell. BikeNet: A mobile sensing system for cyclist experience mapping. *ACM Transactions and Sensor, Networks*, 6(1):6:1–6:39, Jan 2010.

313. B J Fogg. *Persuasive Technology: Using Computers to Change what we Think and Do*. Morgan Kaufmann, 2003.

314. S S Intille. A new research challenge: persuasive technology to motivate healthy aging. *Information Technology in Biomedicine, IEEE Transactions on*, 8(3):235–237, Sept 2004.

315. M Knoll. Diabetes City: How urban game design strategies can help diabetics. In *eHealth'08*, pages 200–204, 2008.

316. Y Lin, J Jessurun, B de Vries, and H Timmermans. Motivate: Towards context-aware recommendation mobile system for healthy living. In *Pervasive Computing Technologies for Healthcare (PervasiveHealth), 2011 5th International Conference on*, pages 250–253, May 2011.

317. R R Fletcher, M Poh, and H Eydgahi. Wearable sensors: Opportunities and challenges for low-cost health care. In *Engineering in Medicine and Biology Society (EMBC), 2010 Annual International Conference of the IEEE*, pages 1763–1766, Sept 2010.

318. B Resch, A Zipf, P Breuss-Schneeweis, E Beinat, and M Boher. Live cities and urban services: A multi-dimensional stress field between technology, innovation and society. In *GEOProcessing 2012 : The Fourth International Conference on Advanced Geographic Information Systems, Applications, and Services*, 2012.

319. J F Coughlin, B Reimer, and B Mehler. Monitoring, managing, and motivating driver safety and well-being. *Pervasive Computing, IEEE*, 10(3):14–21, July-Sept 2011.

320. A Doshi, B T Morris, and M M Trivedi. On-road prediction of driver's intent with multimodal sensory cues. *Pervasive Computing, IEEE*, 10(3):22–34, July-Sept 2011.

321. M Walter, B Eilebrecht, T Wartzek, and S Leonhardt. The smart car seat: personalized monitoring of vital signs in automotive applications. *Personal Ubiquitous Computing*, 15(7):707–715, Oct 2011.

322. R A Katzmann. Transportation policy. In J West, editor, *The Americans with Disabilities Act: From Policy to Practice*, pages 214–237. Milbank Memorial Fund: New York, 1991.

323. J Rosenkvist, R Risser, S Iwarsson, K Wendel, and A Stahl. The challenge of using public transport: Descriptions by people with cognitive functional limitations. *Journal of Transport and Land Use*, pages 65–80.

324. I Audirac. Accessing Transit as Universal Design. *Journal of Planning Literature*, 23(1):4–16, 2008.

325. J Cordeau and G Laporte. The Dial-A-Ride Problem: models and algorithms. *Annals of Operations Research*, 153:29–46, 2007.

326. S Kammoun, M J-M Macé, B Oriola, and C Jouffrais. Toward a better guidance in wearable electronic orientation aids. In *Proc. 13th IFIP TC 13 International Conference on Human-computer interaction - Volume Part IV: Series: INTERACT'11*, pages 624–627, 2011.

327. L Liao, D J Patterson, D Fox, and H Kautz. Learning and inferring transportation routines. *Artificial Intelligence*, 171(5–6):311–331, Apr 2007.

328. P Narasimhan. Assistive embedded technologies. *Computer*, 39(7):85–87, Jul 2006.

329. K Schilling. Robotic and telematic assistant technologies to support aging people. In *Instrumentation, Communications, Information Technology, and Biomedical Engineering (ICICI-BME), 2009 International Conference on*, pages 1–3, Nov 2009.

330. S J Gaukrodger and A Lintott. Augmented reality and applications for assistive technology. In *Proc. 1st International Convention on Rehabilitation engineering; assistive technology: in conjunction with 1st Tan Tock Seng Hospital Neurorehabilitation Meeting, Series: i-CREATe '07*, pages 47–51, 2007.

331. A Hub, J Diepstraten, and T Ertl. Design and development of an indoor navigation and object identification system for the blind. *SIGACCESS Access. Comput.*, (77–78):147–152, Sept 2003.

332. V Kulyukin, C Gharpure, J Nicholson, and S Pavithran. RFID in robot-assisted indoor navigation for the visually impaired. In *Proc. IEEE/RSJ International Conference on Intelligent Robots and Systems, (IROS 2004)*, volume 2, pages 1979–1984 vol. 2, Sept - Oct 2004.

333. S Gehring. Adaptive indoor navigation for the blind. In *GI Jahrestagung (1)'08*, pages 293–294, 2008.

334. Y Chang, S Peng, T Wang, S Chen, Y Chen, and H Chen. Autonomous indoor wayfinding for individuals with cognitive impairments. *Journal of NeuroEngineering and Rehabilitation*, 7(1):45, 2010.

335. H Steg, H Strese, C Loroff, J Hull, and S Schmidt. Europe is facing a demographic challenge ambient assisted living offers solutions. Technical report, March 2006.

336. J Abascal, B Bonail, Á Marco, R Casas, and J L Sevillano. AmbienNet: an intelligent environment to support people with disabilities and elderly people. In *Proc. 10th International ACM SIGACCESS Conference on Computers and Accessibility (Assets '08)*, pages 293–294, 2008.

337. S Ezell. Explaining international IT application leaderhip: Intelligent transportation systems. Technical report, The Information Technology and Innovation Foundation, 2010.

338. GPS.gov. GPS program funding. http://www.gps.gov/policy/funding/, n.d. Access date: 08/10/2012.
339. European Commission Information Society. European broadband: investing in digitally driven growth. http://ec.europa.eu/information_society/activities/broadband/index_en.htm, n.d.
340. Communications Australian Government: Department of Broadband and the Digital Economy. A broadband network for Australias future. http://www.nbn.gov.au/about-the-nbn/what-is-the-nbn/, n.d.
341. National Broadband Plan. National Broadband Plan: Connecting America. Technical report, 2010. Access date: 08/10/2012.
342. European Commission Information Society. European broadband: investing in digitally driven growth. http://ec.europa.eu/information_society/activities/broadband/investment/index_en.htm. Access date: 08/05/2012.
343. D E Boyce. A memoir of the ADVANCE project. *ITS Journal*, 5:103–130, 2002.
344. S Cheon. An overview of Automated Highway Systems (AHS) and the social and institutional challenges they face. Technical report, University of California Transportation Center, 2003.
345. R Herring, A Hofleitner, D Work, O Tossavainen, and A M Bayen. Mobile Millennium - participatory traffic estimation using mobile phones. Technical report, Volvo Research and Educational Research Foundation, n.d.
346. Us, DOT ITS Joint Program Office. The Connected Vehicle Test Bed. http://www.its.dot.gov/connected_vehicle/technology_testbed2.htm, n.d. Access date: 08/15/2012.
347. VICS. How VICS works. http://www.vics.or.jp/english/vics/index.html, n.d. Access date: 08/15/2012.
348. Y Kumagai. Introducing ITS in Asian countries. Technical report, Japan International Transport Institute, 2001.
349. D Crawford. Mixed results for public-private traffic management partnerships. http://www.itsinternational.com/categories/utc/features/mixed-results-for-public-private-traffic-management-partnerships/, n.d. Access date: 08/15/2012.
350. Y Sugawara. Understanding the differences in the development and use of Advanced Traveler Information Systems for Vehicles (ATIS/V) in the Us, Germany and Japan. Technical report, Massachussetts Institute of Technology, 2007.
351. At &t 2009 annual report. http://www.att.com/gen/investor-relations?pid=17393 (Access Date: 08/2012), 2009.
352. D V Dorselaer and J P Breazeale. Sources and uses of financing in the US telecom industry. *Research in Business and Economics Journal*, August 2011.
353. Inc. Google. Google 2009 Annual Report. Technical report, Google, Inc., 2009. Access Date: 08/2012.
354. World Bank. Financing information and communication infrastructure needs in the developing world: Public and private roles. Technical report, World Bank, 2005.
355. I D Constantiou and J Damsgaard. Bundle pricing for Location Based Mobile Services. In *ECIS 2004 Proceedings*, volume 41, 2004.
356. R Ferraro and M Aktihanoglu. *Location-Aware Applications*. Manning Publications, 2011.
357. T Yokota and R J Weiland. ITS system architectures for developing countries: Technical note 5. Technical report, World Bank, 07 2004.
358. Z Belinová, P Bureš and P Jesty. Intelligent Transport System architecture different approaches and future trends. In J Düh, H Hufnagl, E Juritsch, R Pfliegl, H Schimany, and H Schönegger, editors, *Data and Mobility*, volume 81 of *Advances in Intelligent and Soft Computing*, pages 115–125. Springer Berlin/Heidelberg, 2010.
360. Federal Trade Commission. Fair information practice principles. http://www.ftc.gov/reports/privacy3/fairinfo.shtm, n.d. Access date: 08/12/2012.
361. ITS America. ITS America's Fair Information and Privacy Principles. http://www.itsa.org/images/mediacenter/itsaprivacyprinciples.pdf. (Access date: 08/10/2012).

362. CTIA The Wireless Association. Best practices and guidelines for Location-Based Services. Technical report, CTIA, 2010.
363. European Parliament and of the Council. Directive 95/46/ec. *Official Journal of the European, Communities*, pages 31–50, Nov 1995.
364. C Cottrill and P Thakuriah. Protecting location privacy. policy evaluation. *Transportation Research Record, Journal of the Transportation Research Board*, 2215:67–74, 2011.
365. C Cottrill. *An Analysis of Privacy in Intelligent Transportation Systems and Location Based Services: Policy, Technology and Personal Preferences*. Dissertation, University of Illlinois at Chicago, 2011.
366. Wireless Telecommunications Bureau. Location-Based Services: An overview of opportunities and other considerations. Technical report, Federal Communications Commission, May 2012.
367. Federal Communications Commission. Amendment of the commissions rules regarding dedicated short-range communication services in the 5.850-5.925 ghz band (5.9 ghz band). Technical Report FCC 03-324.
368. European Commission. Directive 671: harmonised use of radio spectrum in the 5 875-5 905 mhz frequency, 08 2008. Access date: 08/01/2012.
369. Austroads. Intelligent Vehicles and Infrastructure: The case for securing 5.9ghz. Technical Report AP-R330/08, AUSTROADS Inc., 2008.
370. Vehicle Infrastructure Integration Consortium (VIIC). Cooperative Transportation Systems and the OpenPlatform Concept. http://downloads.transportation.org/Executive_Leadership_ Team/8%20-%20VIIC%20Perspectives%20on%20the%20Open%20Platform%20Concept %20-%20Final.pdf, 2008. Access date: 08/01/2012.
371. Federal Register. Visual-manual NHTSA driver distraction guidelines for in-vehicle electronic devices; notice. Technical report, Feb 2012.
372. G Kuk and M Janssen. The business models and information architectures of smart cities. *Journal of Urban Technology*, 18(2):39–52, 2011.
373. The White House. Open Government Directive. http://www.whitehouse.gov/sites/ default/files/omb/assets/memoranda_2010/m10-06.pdf, 2009. Access date: 08/05/2012.
374. Australian Government Department of Finance and Deregulation. Declaration of Open Government. http://agimo.govspace.gov.au/2010/07/16/declaration-of-open-government/, n.d. Access date: 08/01/2012.
375. European Commission. Open data: An engine for innovation, growth and transparent governance. Technical report, European Commission, 2011.
376. HM Government. Putting the frontline first: Smarter Government. Technical report, 2009. Access date: 08/05/2012.
377. N Huijboom and T van den Broek. Open data: an international comparison of strategies. *European Journal of ePractice*, 12, March/April 2011.
378. White House. Obama Administration announces Big Data initiative. http://www.whitehouse. gov/sites/default/files/microsites/ostp/big_data_press_release.pdf, 2012. Access date: 08/10/ 2012.
379. McKinsey Global Institute. Big data: The next frontier for innovation, competition, and productivity. Technical report, McKinsey Global Institute, June 2011.
380. D DiBiase, M DeMers, A Johnson, K Kemp, A T Luck, B Plewe, and E Wentz. Cartography and Geographic Information Science. *Cartography and Geographic Information Science*, 34(2):113–120, 2007.
381. S Winter, M Sester, O Wolfson, and G Geers. Towards a Computational Transportation Science. *SIGMOD Rec.*, 39(3):27–32, February 2011.
382. University of Illinois at Chicago. IGERT: Phd program in computational transportation science. http://cts.cs.uic.edu/, n.d.
383. H Chourabi, T Nam, S Walker, J R Gil-Garca, S Mellouli, K Nahon, T A Pardo, and H S Scholl. Understanding smart cities: An integrative framework. In *HICSS*, pages 2289–2297, 2012.

384. W Lam. Barriers to e-government integration. *The journal of Enterprise Information Management*, 18:511–530, 2005.

385. J R Gil-Garca and T A Pardo. E-government success factors: Mapping practical tools to theoretical foundations. *Government Information Quarterly*, 22:187–216, 2005.

386. S R Arnstein. A ladder of citizen participation. *Journal of the American Institute of Planners*, 35(4):216–224, 1969.

387. Manchester City Council. Smart metropolitan areas realised through innovation and people [smart]iP. Technical report, 2010. Access date:08/10/2012.

388. W Boger. The harmonization of european products liability law. *Fordham International Law Journal*, 7:1–60, 1983.

389. P Elliott and E Stanley. Liability issues with Intelligent Transport Systems, 2 2010. Access date: 08/01/2012.

390. C J Biederman and D Andrews. Applying Copyright Law to User-Generated Content. *Los Angeles Lawyer*, pages 12–18, 5 2008.

391. R Latham, C C Butzer, and J T Brown. Legal Implications of User-Generated Content: YouTube, MySpace, Facebook. *Intellectual Property and Technology Law Journal*, 20(5):1–11, 2008.

392. S Holmes and P Ganley. User-Generated Content and the law. *Journal of Intellectual Property Law & Practice*, 2(5):338–344, 2007.

393. L Van Wel and L Royakkers. Ethical issues in web data mining. *Ethics and Inf. Technol.*, 6(2):129–140, June 2004.

394. A Vedder. In R Spinello and H Tavani, editors, *Readings in CyberEthics*, chapter KDD, Privacy, Individuality, and Fairness, pages 404–412. Jones and Bartlett, 2001.

395. American Institute of Certified Public Accountants Inc and Canadian Institute of Chartered Accountants. Trust services principles, illustrations and criteria, 2009. Access date: 08/01/ 2012.

396. H Xu, H Teo, and B C Y Tan. Predicting the adoption of Location-Based Services: The role of trust and perceived privacy risk. In *Proc. International Conference on Information System (ICIS '05)*, number 71, 2005.

397. J D Gregory. Solving legal issues in electronic government: Jurisdiction, regulation, governance. *Canadian Journal of Law and Technology*, 1(3):1–26, 2002.

398. M S Ribble, G D Bailey, and T W Ross. Digital citizenship: Addressing appropriate technology behavior. *International Society for Technology in Education*, pages 6–12, Sept 2004.

399. L Palen. Mobile telephony in a connected life. *Communications of the ACM - Robots: intelligence, versatility, adaptivity*, 45(3):78–82, Mar 2002.

400. J E Katz and R E Rice. *Social Consequences of Internet Use: Access, Involvement, and Interaction*. MIT Press, 2002.

401. Y Eshet-Alkalai. Digital literacy: A conceptual framework for survival skills in the digital era. *Journal of Educational Multimedia and Hypermedia*, 13(1):93–106, Jan 2004.

402. K L Hacker and J V (eds) Dijk. *Digital Democracy: Issues of Theory and Practice*. Sage, 2000.

403. P DiMaggio and E Hargittai. From the "digital divide" to "digital inequality": Studying internet use as penetration increases. 2001. Access Date: 10/2012.

404. N R Velaga, M Beecroft, J D Nelson, D Corsar, and P Edwards. Transport poverty meets the digital divide: accessibility and connectivity in rural communities. *Journal of Transport Geography*, 21(0):102–112, 2012.

405. K Brown, S W. Campbell, and R Ling. Mobile phones bridging the digital divide for teens in the us? *Future Internet*, 3(2):144–158, 2011.

406. D Radovanovic. Digital divide and social media: Connectivity doesnt end the digital divide, skills do. *Scientific American*, December 2011.

407. P L Mokhtarian. If telecommunication is such a good substitute for travel, why does congestion continue to get worse? *Transportation Letters: The International Journal of, Transportation Research*, pages 1–17, 2009.

408. S Bamberg, I Ajzen, and P Schmidt. Choice of travel mode in the theory of planned behavior: The roles of past behavior, habit, and reasoned action. *Basic & Applied Social Psychology*, 25(3):175, 2003.

409. C Domarchi, A Tudela, and A González. Effect of attitudes, habit and affective appraisal on mode choice: an application to university workers. *Transportation*, 35:585–599, 2008.

410. L Eriksson, J Garvill, and A M Nordlund. Interrupting habitual car use: The importance of car habit strength and moral motivation for personal car use reduction. *Transportation Research Part F: Traffic Psychology and Behaviour*, 11(1):10–23, 2008.

411. L. Tang and P. Thakuriah. Will psychological effects of real-time transit information systems lead to ridership gain? *Transportation Research Record, Journal of the Transportation Research Board*, 2216:67–74, 2011.

412. L Tang and P Thakuriah. Ridership effects of real-time bus information system: A case study in the city of chicago. *Transportation Research Part C: Emerging Technologies*, 22(0):146–161, 2012.

413. P Thakuriah, L Tang, and W Vassilakis. Spatio-temporal effects of bus arrival time information. In *Proc. 4th ACM SIGSPATIAL International Workshop on Computational Transportation Science*, CTS '11, pages 6–11, 2011.

414. N Tilahun, P Thakuriah, and Y Mallon-Keita. Factors determining transit access by car-owners: Implications for intermodal passenger transportation planning. In *Proc. Transportation Research Board 2013 Annual Conf.*, 2013.

415. J Palfrey and U Gasser. *Born Digital: Understanding the First Generation of Digital Natives*. Basic Books, Inc., 2008.

416. K Zickuhr. Generations 2010. Technical report, Pew Research Center, 12 2010. Access Date: 08/2012.

417. P. Thakuriah, S Menchu, and L Tang. Car-ownership among young adults: Generational and period-specific perspectives. *Transportation Research Record*, 2156:1–8, 2010.

418. M Sivak and B Schoettle. Recent changes in the age composition of us drivers: Implications for the extent, safety, and environmental consequences of personal transportation. *Traffic Injury Prevention*, 12(6):588–592, 2011.

419. Allstate. Economy puts the brakes on parents spending for teen cars driving expenses. http://www.allstatenewsroom.com/channels/News-Releases/releases/economy-puts-the-brakes-on-parents-spending-for-teen-cars-driving-expenses, 2011. Access Date: 08/2012.

420. P R Stopher and C (ed.) Stecher. *Travel Survey Methods: Quality and Future Directions*. Elsevier, 2006.

421. J Wolf. Application of new technologies in travel surveys. In P R Stopher and C Stecher, editors, *Travel Survey Methods: Quality and Future Directions*, pages 531–541. Elsevier, 2006.

422. M. Nanni, R. Trasarti, C. Renso, F. Giannotti, and D. Pedreschi. Advanced knowledge discovery on movement data with the GeoPKDD system. In *Proc. 13th International Conference on Extending Database Technology (EDBT '10)*, pages 693–696, 2010.

423. F Giannotti, M Nanni, D Pedreschi, F Pinelli, C Renso, S Rinzivillo, and R Trasarti. Mobility data mining: discovering movement patterns from trajectory data. In *Proc. 2nd International Workshop on Computational Transportation Science (IWCTS '10)*, pages 7–10, 2010.

424. C Bhat and F Koppelman. Activity-based modeling of travel demand. In R W Hall, editor, *Handbook of Transportation Science, volume 56 of International Series in Operations Research & Management Science*, pages 39–65. Springer US, 2003.

425. Y Shiftan and M Ben-Akiva. A practical policy-sensitive, activity-based, travel-demand model. *The Annals of Regional Science*, 47:517–541, 2011.

426. Y Zhang, P Stopher, and Q Jiang. Developing tour-based data from multi-day gps data. In *Australasian Transport Research Forum (ATRF)*, 2010.

427. J Auld and A Mohammadian. Activity planning processes in the adapts activity-based modeling framework. In *Proc. International Conference of the International Association for Travel Behaviour Research*, 2009.

428. P Zhang. Toward a positive design theory: Principles for designing motivating information and communication technology. In M Avital, R J Boland, and D L Cooperrider, editors, *Designing Information and Organizations with a Positive Lens (Advances in Appreciative Inquiry, Volume 2)*, pages 45–74. Emerald Group Publishing Limited, 2007.

429. D Te'eni, J M Carey, and P Zhang. *Human-Computer Interaction: Developing Effective Organizational Information Systems*. Wiley, 2006.

Index

P. (Vonu) Thakuriah and D. G. Geers, *Transportation and Information*, SpringerBriefs   125
in Computer Science, DOI: 10.1007/978-1-4614-7129-5, © The Author(s) 2013